PETROLEUM ECONOMICS AND OFFSHORE MINING LEGISLATION

PETROLEUM ECONOMICS
AND
OFFSHORE MINING LEGISLATION

A geological evaluation

ANTON PEDRO HENDRIK VAN MEURS

Geological Institute State University, Utrecht
The Netherlands

ELSEVIER PUBLISHING COMPANY
Amsterdam-London-New York
1971

ELSEVIER PUBLISHING COMPANY
335 Jan van Galenstraat
P.O. Box 211, Amsterdam, The Netherlands

ELSEVIER PUBLISHING CO. LTD.
Barking, Essex, England

AMERICAN ELSEVIER PUBLISHING COMPANY, INC.
52 Vanderbilt Avenue
New York, New York 10017

LIBRARY OF CONGRESS CARD NUMBER: 76-135486

ISBN 0-444-40889-4

WITH 42 ILLUSTRATIONS AND 19 TABLES

PRINTED IN THE NETHERLANDS

Preface

Introduction

Our present highly technical and sophisticated civilization demands a continuous supply of minerals. Minerals are required to produce most of the goods which together constitute material welfare; further, primary-energy production is based almost entirely on coal, oil, natural gas and uranium.

Minerals are unevenly distributed throughout the world in the earth's crust. Larger states can produce an adequate supply of these minerals domestically, while other states are forced to import large quantities. International trade in minerals is understandably an important factor in the world's political, social and economic situations. The mineral policy of individual countries, therefore, has international repercussions.

The international aspects of mineral production, plus its tremendous internal importance, combine to make new mining legislation within a country of vital political significance. It is not surprising that considerable political activity in western Europe followed the discovery of the Slochteren gas field in The Netherlands. The eye of the entire oil industry was fixed upon the North Sea area at the time. The states surrounding the North Sea consequently prepared offshore-mining legislation. The Dutch legislation in particular was subject to extensive discussion, with four successive Dutch governments taking part in its formulation. Problems of a mixed geological, technical, juridical and economical character needed to be solved.

These problems inspired the author to evolve the main lines of his thesis. This study is of a theoretical character, but the extensive discussions preceding the establishment of the offshore-mining legislation in The Netherlands make a further detailed deliniation of the thinking behind such legislation appear practically useful.

About this book

This book consists of nine chapters. The character of each chapter varies considerably. To aid the student of this thesis in finding the subjects which most interest him, the author will here outline the book's contents.

Chapter I lays out the scope of the study and gives some elementary data about minerals and the petroleum industry. The forementioned data are of little interest to anyone who is familiar with the industry.

Chapter II reviews basic information about the estimation and distribution of oil and gas reserves. The only original contribution by the author is the definition of the economically-recoverable reserves. This chapter is likewise of little interest for petroleum engineers and geologists.

Chapter III compiles the essential information about oil and gas markets. The author has attempted to evaluate this information from a geologist's point of view. Although most of the information will be familiar to a petroleum economist, he will discover some original and personal viewpoints. The second part of the chapter is an introduction to petroleum law.

Chapter IV deals with investment analysis in petroleum exploration and production. This is a rather specialized subject. The author evaluated particularly the handling of geological risk.

Chapter V is an extension of Chapter IV. It gives special weight to the influence of mining legislation on investment analysis by petroleum companies.

Chapter VI forms the core of this work. It is the first published attempt at analyzing the essential building stones of the financial arrangements in a petroleum law from a geological point of view—that is, based on the general characteristics of a national reserve of petroleum.

Chapter VII demonstrates the practical use of the analysis of Chapter VI by examining the discussions that surrounded the construction of the complicated provisions of the Dutch offshore-mining law. Those familiar with these discussions will probably be shocked by the conclusions of this thesis.

Chapter VIII is a limited study of petroleum legislation in different industrialized countries. Conclusions reached in Chapter VI are applied in this survey.

Chapter IX is a comparable study of petroleum legislation in the Middle East.

Acknowledgements

It is a pleasant duty to acknowledge those who were involved in the preparation of this thesis and in my professional training.

In particular I am indebted to Prof. Dr. Ir. R. W. van Bemmelen and Prof. Dr. A. I. Diepenhorst, who promoted and guided this thesis. Their criticism and suggestions were of great importance to this study.

Prof. Dr. M. G. Rutten, Prof. Dr. G. H. R. von Koenigswald, Prof. Dr. J. Wemelsfelder, Prof. Dr. D. J. Doeglas, and Prof. Mr. P. Schierbeek, assisted in my professional training. I feel indebted for their valuable courses.

Grateful acknowledgement is made to Ir. J. J. Arps, Drs. J. N. F. Bakker, Ir. C. J. A. Berding, Ir. J. Bijl, Drs. G. C. Brouwer, Dr. H. Dekker, Mr. J. Jonkers, Drs. P. Krens, Dr. P. Nieuwendorp, Ir. F. L. van Berckel, and Dr. J. Weeda for their assistance, criticism, and suggestions in the earlier and later stages of the preparation of my thesis.

I like to express my gratitude to Miss Kelley for correcting the English of the manuscript.

The assistance of mr. A. J. van Poppen, mr. F. J. L. Henzen, mr. I. M. Santoe, and mr. J. H. Smits, with typework, the drawing of the illustrations and the technical preparation of the manuscript, was greatly appreciated.

This thesis is completed with the generous financial assistance of The Netherlands Organization for the Advancement of Pure Research and a subsidy of the Utrechts Universiteitsfonds.

The author is solely responsible for the opinions expressed in this thesis.

Contents

Chapter I

Introduction

Scope of the study

Petroleum companies need to explore for petroleum to assure themselves of continuous possibilities for production. When they enter an unexplored area, they are reluctant to make significant payments to government for exploration and eventual production rights. Risks are high and exploration is expensive. The companies are not interested in marginal fields, but are geared toward exploring for a rich discovery only. Such a "bonanza" will pay for many "dry holes", and assure a future share in the market.

Most governments welcome exploration by petroleum companies within their jurisdictional borders. The production of petroleum may lead toward increased material welfare. If a rich field is discovered, however, the operator is not only making a normal profit but is also collecting a high rent. This rent may go outside the state—especially in the case of small or developing countries—while these same funds could have been used for the growth of the national economy. Governments, therefore, like to assure themselves of a large share in these rents.

Petroleum legislation must be constructed to induce companies to take risks but to avoid at the same time unnecessary losses to the economy. What should the framework of such a legislation be? That is the scope of the present study.

The study develops in four steps. First, some general aspects of minerals, the petroleum industry, petroleum reserves, and finally the oil and gas markets with their juridical and political problems are examined (Chapters I—III). Secondly, the decision-making process in petroleum companies and the influence of mining legislation on this process are studied (Chapters IV and V). Thirdly, the conditions for a successful mining legislation are analyzed (Chapter VI). And finally, these conditions will be compared with actual petroleum laws—with special reference to offshore-mining legislation.

The author is convinced of the fact that petroleum companies and governments are mostly fortunate in having highly experienced experts to deal with those problems—which are generally too complicated to be handled by a single person, as can be seen in Fig. 1. This study is, therefore, necessarily restricted to an elaboration of some personal ideas.

Minerals

The term "mineral" can be defined in numerous different ways. "Mineral" will here be used in the context of its economic importance; thus an "economic definition", such

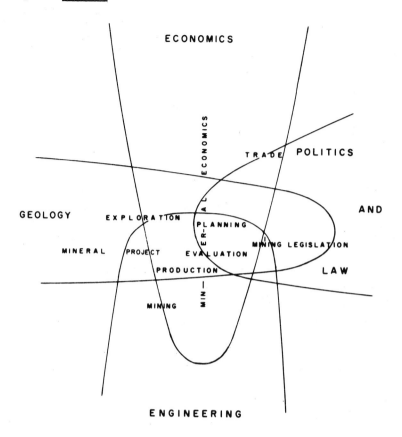

ECONOMICS

MINERAL ECONOMICS

TRADE POLITICS

GEOLOGY

AND

EXPLORATION

PLANNING

MINING LEGISLATION

MINERAL

PROJECT

EVALUATION

PRODUCTION

LAW

MINING

MINING

ENGINEERING

Fig. 1. Mineral economics and related sciences

as the following is here functional: A mineral is any of the various naturally occuring homogeneous or apparently homogeneous, and usually, but not necessarily, solid substances obtained for man's use in the main from the ground, in such a way that natural replenishment is not possible while a price can be established for each substance.

This definition includes all those minerals which are traded internationally and nationally, such as coal, iron ore, borax, clay, salt, precious stones, oil, helium, etc. It excludes those substances that are produced synthetically or those naturally-occurring substances that are of no use to mankind. Other useful substances, as rain water, are excluded as soon as they are excluded from trade—and when a price cannot be established.

Minerals play a vital role in our world. They compose a large percentage of total trade in commodities. Table 1 lists those states where minerals occupy a central place in exports. In 33 countries minerals contribute 25% or more of the total value of exports. With the exception of South Africa, all of these 33 are developing countries.

Minerals form the bulk of the commodities shipped by sea transport. Since 1960

TABLE I

LIST OF COUNTRIES IN WHICH THE EXPORT OF MINERALS IS OF GREAT RELATIVE IMPORTANCE[1]

Countries of which 50% or more of the value of the export is formed by minerals:		Countries of which 25 – 50% of the value of the exports is formed by minerals:	
Algeria	– petroleum	Cyprus	– different minerals
Bolivia	– tin	Indonesia	– petroleum, tin
Brunei	– petroleum	Jordan	– phosphates
Central African		Malaysia	– tin
Republic	– diamonds	Marocco	– phosphates
Chili	– copper	Nigeria	– petroleum, tin
Congo	– copper	Peru	– copper
Gabon	– manganese,	Rwanda	– different minerals
	uranium,	South Africa	– gold, diamonds
	petroleum	Togo	– phosphates
Guyana	– bauxite,	Tunisia	– phosphates
	manganese,	Rhodesia	– asbestos
	diamonds		
Iran	– petroleum		
Iraq	– petroleum		
Jamaica	– bauxite		
Kuwait	– petroleum		
Liberia	– iron ore		
Lybia	– petroleum		
Mauretania	– iron ore		
Saudi Arabia	– petroleum		
Sierra Leone	– diamonds,		
	bauxite		
Surinam	– bauxite		
Trinidad	– petroleum		
Venezuela	– petroleum		
Zambia	– copper		

[1] Source: Yearbook of International Trade Statistics, 1966. United Nations

more than 50% of all shipments consists of oil, while iron and manganese, coal, bauxite, alumina and phosphate contribute another 17%.

In industrialized countries, minerals are essential to the promotion of material welfare. High standards of living are the product of diversified industrial systems, which require a great variety of minerals for maintenance and growth. A car cannot be built without bauxites, iron ores, copper ores, lead ores, tungsten ores, etc., while it is impossible to drive the car without petroleum.

LOVERING (1943) gives an excellent description of the role of minerals in his book "Minerals in World Affairs". When minerals enter economic life, they are characterized by a number of common features. According to Lovering, features peculiar to mineral economics can be summarized under four points:

(1) *Localized occurrence.* Mineral resources are distributed unevenly throughout the earth's crust. For instance, about 75% of world oil reserves is presently found in the Middle East and North Africa (cf. HILL and EALES, 1970), while western Europe and the U.S.A. consume 60% of the world production of crude oil. Thus the

location of oil's natural occurrence determines to a great extent the pattern of the world's oil trade; moreover it is the foundation of the international oil policy of petroleum-consuming and producing countries. The same generalization applies to almost all other minerals. Even sand and gravel can be subject to an intensive trade. The downstream traffic on the Rhine crossing the German-Dutch border includes yearly about 18×10^6 tons of sand and gravel. This figure shows that even sand and gravel are materials which can be profitably exported. Another aspect of localized occurrence is that the area in which there are underlying minerals is extremely small in comparison with the total land surface. For instance, the total acreage of the U.S.A. oil and gas producing fields is about 2% of the area considered favorable for petroleum occurrence. A number of the most important fields in the U.S.A. cover an area of less than two square miles (5 km^2).

(2) *Exhaustibility*. The economically-recoverable reserves of most mineral deposits are limited. In particular the so-called strategic minerals, such as oil, copper, lead, zinc, gold, etc. are scarce. If a deposit is mined for a certain length of time, it becomes exhausted. This is a central problem for the mineral producer because the investments necessary to begin and continue operations elsewhere must be recovered before the deposit is exhausted. Additionally, new profitable deposits must be discovered. Governments of mineral-exporting countries in particular must face the problem of ultimate exhaustion of this source of welfare. This problem is basic to all the countries listed in Table I, and is a continuous source of tension between these countries and the mining or oil companies.

(3) *Increase of costs* with depth and geographical expansion of activity. For instance, the average depth of the new oilfield wildcats in the U.S.A. was during 1946 about 4,000 ft., while the same figure for 1969 was nearly 5,500 ft. The deeper the production, the greater the cost. Petroleum exploration has spread since World War II to all corners of the world. At present exploration is successfully being carried on at the North Slope of Alaska, where drilling is about five times more costly than is normal in other parts of the U.S.A. An important new development is the exploration of the continental shelf, where drilling is far more expensive than in onshore activities. Operations on the continental shelf were going on in four countries (Venezuela, Saudi Arabia, U.S.A. and U.S.S.R.) in 1954, while in 1968 such activities had started or continued in 43 countries (GIBSON, 1969). Increases in costs can sometimes be reduced by the application of new techniques and better management. In a report of the NATIONAL PETROLEUM COUNCIL (1967) the cost reductions due to technical development are estimated to amount in fifteen years (1950 - 1965) in the U.S.A. up to U.S.$ 1.00 per barrel.

(4) *Discovery hazards*. New minerals must be discovered to replace other minerals that are produced. This puts a constant burden on the mining industry and is source of uncertainty about the future. No mining company, of course, can predict whether or not its exploration projects will succeed. Most minerals are difficult to find; the petroleum industry, for example, is virtually distinguished by its high level of exploratory activity directed toward the discovery of new oil reserves. Since World War II approximately 11% of the new field wildcats have been successful in the U.S.A. In addition, no company can guess what other companies will discover. It is

possible that highly successful exploration by certain companies may change the entire economic scene for a particular mineral. This has perhaps been the result in the Alaska oil discoveries and the iron-ore deposits recently found on the north-west coast of Australia. To rephrase the problem—the success of one company may contribute to the failure of others. LOVERING (1943) gives these situations two different titles: the undersupply hazard and the oversupply hazard. Such uncertainties about the future are also most important to the producing countries. Considerable discoveries may lower prices, perhaps negatively influencing the value of exports and the government's bargaining position with regard to the companies. The political oil situation has changed drastically in the Middle East with the rise of the new oil-producing countries in Africa: Lybia and Nigeria.

The four previously mentioned factors characterize most minerals, but are especially typical of oil and gas exploration and production. This study focuses upon the petroleum industry.

The oil and gas industry

Before turning to some specific problems, the author will describe some general features of the oil and gas industry. Particular attention will be drawn to the geology of oil and gas, and to their exploration, production and transportation.

Oil and Gas

Oil and gas are naturally occuring mixtures of hydrocarbons and other organic compounds containing small amounts of oxygen, nitrogen, sulfur and sometimes metals. The boilingpoint of all the different constituents varies widely. For instance, natural gas and oil may contain quantities of methane (CH_4) with a boiling point of $-161°$ C, while other compounds have boiling points above $450°$ C. Under surface conditions, oil is a liquid and natural gas is a gas, but between these two exists an intermediary group called "natural-gas liquids". The components of this group are liquid hydrocarbons consisting primarily of light and intermediary hydrocarbons and occurring as a free gas phase or in solution with crude oil in an oil reservoir. Natural gasoline, condensate and liquified-petroleum gases fall into this category (see Appendix IV).

Oil and gas are thought to develop from organic material. Some substances from animal or plant debris, enclosed in sediments, are transformed at a certain depth within the earth's crust to oil and gas. Before an economically-recoverable accumulation of oil or gas can be formed a specific set of conditions must exist.

The oil or gas must first be removed from the sourcerock and transported through the groundwater by the buoyancy or the hydrodynamics of the groundwater system. Secondly, the oil or gas must accumulate in a reservoir. A reservoir can be a porous sandstone or any other rock containing sufficient pores to hold the oil. The oil or gas must be prevented from escaping to the surface of the earth by an impermeable rock such as a shale or a salt layer. Thirdly, the relation between the permeable and impermeable rock must be such that the oil or gas are trapped under the impermeable rock while the hydrodynamics of the groundwater are unable to remove the oil or the gas from the trap. Fourthly, to be recoverable the porous rock must be sufficiently permeable to allow the oil or gas to flow to a well. And finally, the accumulation must be large enough to permit profitable production.

Roughly two different types of traps can be distinguished. The first, structural trap, is formed by geological movements in the earth's crust, resulting in folds and faults. The most common type is the anticlinal trap, where oil or gas is trapped beneath a dome-shaped impermeable layer in a reservoir. Some of the Middle East oil fields are examples of this group. The formation of the second type—the stratigraphic trap—is controlled by the manner in which the sediments are deposited. Important stratigraphic traps are porous fossil reefs.

The relation between oil and gas is different in various pools. Gas may be dissolved in oil. The oil may be undersaturated with gas or saturated. In the latter case a gas cap can be found above the oil. Gas may also be found with no relation to nearby oil (see Appendix IV).

Exploration

Profitable accumulations of oil and gas are difficult to find; thus a wide variety of exploratory techniques has been developed to seek them out. The choice among the various techniques depends upon many factors, including geographical conditions, the expected depth of the reservoir, the expected trap and the economic conditions. If an entirely new area must be explored, a certain sequence of phases in exploration can be distinguished. In the beginning, general surveys are performed which cover vast territories at a low cost. Such surveys seek to generate a general insight of the area. Photogeology, gravimetry and magnetometry are frequently applied. Photogeology employs air reconnaissance, and is best adapted to onshore problems where the surface of the earth's crust gives an insight into the geology of deeper formations. Magnetometric surveys also use aircraft and are applicable offshore as well as onshore. These measurements give information about the broad outlines of geological provinces and the expected sediment thicknesses.

As soon as some general knowledge has been obtained about an area, more specific information can be obtained onshore with geological mapping by field crews. The expense of this work is within reason, and if geological conditions are suitable, such surveys may reveal much information about the possible presence of traps and reservoir rocks. Offshore, seismic work is generally the second step. Seismic measurements contribute considerably to the discovery of potential structural traps. Furthermore, the cost of seismic work offshore is usually not prohibitive. Onshore seismic work is limited to those areas where important structural traps are expected.

The last phase in the exploration is drilling. The drilling of a new field exploration well (wildcat) is expensive, especially offshore where floating drilling rigs must be used. Only an exploration well can prove conclusively the existence of an oil and/or gas field. Apart from rock samples, new well-logging techniques can add much to existing knowledge about the formations beneath the earth's crust. Most well-logging techniques consist of the continuous measurement of physical characteristics of the well rock of the bore hole—such as resistivity, natural and induced radioactivity, spontaneous potential, etc. Apart from well logging, various other tests are necessary to know the volume of oil or gas which a well can produce after their presence has been established. The most important is the production test.

The succession of the different exploration methods can be studied in Fig. 2, which illustrates exploratory activity in Lybia in the ten years after 1956. The search for oil and gas beneath the continental shelf is characterized by a first peak of gravimetric surveys, followed by intensive seismic work, with, finally, drilling needed to prove the presence of productive reservoirs.

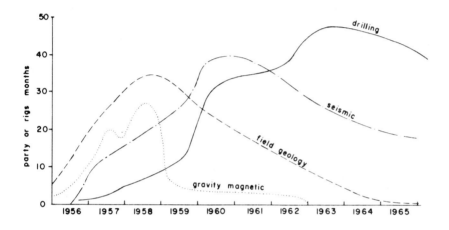

Fig. 2. The sequence of exploration activities in Lybia. Source: *Bull. Am. Assoc. Petrol. Geologists,* 1966. p.1696.

In the first years of exploration mainly structural traps are sought. These traps can be recognized early from detailed geologic or seismic work. The search after stratigraphic traps generally follows rapidly upon evidence that such a trap may be present. Stratigraphic traps can be of considerable importance; thus detailed modern stratigraphic work may lead to significant discoveries after sufficient drilling has been done to collect a general stratigraphic picture in the geologic province. Apart from the above-mentioned methods, geochemical studies may assist in the discovery of oil and gas.

The main developments in exploration techniques are taking place at present in the fields of seismology, modern stratigraphy, well logging and geochemistry, while the computer is assisting more and more in the processing of data.

Development

As soon as a field has been located by an exploration well, work is begun to delineate it. One well alone gives only partial information about a field; a number of wells is needed to fix the extent of the field. Wells drilled to seek this information are called step-out wells.

Following the conclusion that a field can be profitably produced, sufficient development or production wells are drilled to allow a profitable flow of oil or gas and surface equipment is installed. Surface equipment includes the "christmas trees", all necessary pipelines and stock tanks, and equipment to clean gas or oil and to separate the oil, natural-gas liquids, gas, and water. Offshore, all these installations must be concentrated on a production platform, which is a highly technologically developed piece of equipment.

Production

Two essentially different recovery techniques exist to produce oil or gas: that is, primary and secondary-recovery techniques. The primary recovery includes the oil and gas obtainable only through the proper utilization of the natural-reservoir energy; while secondary recovery of oil and gas is made possible by artificially supplying energy to the reservoir.

The natural-reservoir energy may be supplied by various combinations of sources. For example, the water beneath the oil or gas in the reservoir may be under hydrostatic pressure. As soon as a well has penetrated the reservoir and the pressure is reduced, the oil or gas will be forced by this pressure towards the well bore resulting in a flow of oil or gas into the bore hole. Or the expansion of a gas cap above the oil, together with the solution gas in the oil, may supply the needed energy. If only gas is present, its expansion or a water drive may provide the natural energy of the reservoir. CLEGG (1967) calculated that about 27% of the world-oil reserve can be recovered through these primary-recovery techniques. This means that without additional effort, about 73% of the oil will remain in the earth's crust. Of the different drive mechanisms for oil, the water drive gives the best recoveries, sometimes as high as 85% (ARPS et al., 1967). Primary-recovery rates for gas are generally much higher—usually about 75%.

Secondary-recovery techniques consist mainly of water and gas injection. By pumping water into the reservoir beneath the oil, a type of artificial water drive can be established. The injected water forces the oil towards the well. A type of artificial gas-cap drive or solution-gas drive can be constructed if gas that is produced along with the oil (or from another reservoir) is pumped back into the reservoir. This increases the gas pressure resulting in greater oil recovery.

Secondary recovery requires additional investments, but since the costs of exploration in new areas and after deeper reservoirs increase, the attractiveness of investing in secondary-recovery equipment increases commensurably. Originally the secondary-recovery investments were made when a field was almost at the economic limits of its primary production. Presently the secondary-recovery measures are initiated at an early stage in production, which may add to the final total oil recovery. Apart from the economic considerations, the decision to install secondary-recovery equipment depends on the reservoir characteristics. Apart from water and gas injection, thermal methods are applied to low gravity oil reservoirs. The increase in reservoir temperature reduces the viscosity of the oil and hence increases the oil mobility.

An interesting development at present is the use of nuclear explosions to increase the recovery of highly viscous oil; however, the utilization of these explosions is still in an experimental phase. On an smaller scale, techniques to enlarge the permeability around the bore hole, such as formation fracturing and acidizing, are frequently applied.

During production the equipment is heavily attacked by corrosion and other events that diminish the production capacity, making repairs and renewals of equipment in the well necessary. Such activities are called workovers.

When the oil reaches the surface it must be treated to remove water from the crude oil and to separate oil, natural-gas liquids and gas. This is done in separators. Oil and natural-gas liquids can be stocked for a reasonably short time. Gas is normally transported immediately after it has been cleaned of undesirable substances such as hydrogen sulphide or carbon dioxide. This may contribute to a considerable sulfur production along with the gas.

Transportation

Natural gas usually is transported through a pipeline. At present, however, transportation over long distances is possible with tankers specially designed to transport natural gas. For instance, huge gas transports are planned with such tankers between Brunei in Southeast Asia and Japan.

Oil can also be transported through pipelines, a method of transportation which is developing rapidly. Transport costs are low through modern large-diameter, thin-wall pipelines. Compressor stations are largely automated. An important development with oil as well as gas has been the use of giant tankers. A crew of only thirty men is able to manage the transport of 200,000 tons of oil. The use of these tankers becomes even more efficient by the construction of remote loading and unloading facilities. These advances favor the development of offshore oil production, where oil can be produced, stored, and moved into the tankers without using onshore installations.

Literature

Arps, J. J., Brons, F., van Everdingen, A. F., Buchwald, R. W. and Smith, A. E., 1967. A statistical study of recovery efficiency. *Am. Petr. Inst.*, D 14: 1-33.

Clegg, M. W., 1967. Secondary recovery methods—today and tomorrow. *World Petrol.*, 38(4) : 82-86.

Gibson, R., 1969. The spreading offshore search. *World Petrol.*, 40(2): 26-27.

Hill, R. and Eales, R., 1970. Where the world oil industry is heading. *Intern. Management*, 25(1): 21-26.

Lovering, T. S., 1943. *Minerals in World Affairs*. Prentice Hall, Englewood Cliffs, N. J., 394 pp.

National Petroleum Council, 1967. *Impact of New Technology on the U.S. Petroleum Industry 1945-1965*. Natl. Petrol. Council, Washington, D.C., 341 pp.

Chapter II

Oil and gas reserves

Introduction

The quantities of oil and gas believed to be or actually available in a new mining-bill's area of jurisdiction play an important role in the manner mining legislation is constructed and in the course of the negotiations between government and oil companies.

Two aspects are of prime importance: the total amount of economically-recoverable oil and gas, and the distribution of the total reserves in various pools and fields.

For instance, if a new area opened up for concessions appears to the companies to be an attractive one for exploration, due to large expected reserves, the government's bargaining position is strong.

The distribution of field characteristics is, however, of great importance. For instance, the area may contain a substantial total reserve of oil and gas in one large and profitable field, but this reserve can also be distributed in numerous marginal and submarginal fields. Mining legislation must be radically different in each case for a state to maximize profits from the minerals within its borders.

There are considerable uncertainties, especially during the early days of exploration in an entirely new region. The understanding of and dealing with these uncertainties is of crucial importance to an optimal strategy for both industry and government. Chapter IV discusses this question extensively.

In this chapter three different subjects will be discussed in more detail:

(1) The calculation of a reserve for a single pool.
(2) The calculation of a reserve beneath a large area encompassing several pools.
(3) The distribution of several characteristics of the different pools.

Calculation of the reserve for a single pool

The calculation of the reserve for a single oil or gas pool is based on various techniques described in works on oil evaluation such as those of CAMPBELL (1959), FRICK (1962) and HUGHES (1967) or in numerous articles of which can be cited: ARPS (1956), ARPS (1945) or VAN DER LAAN (1968). A brief summary of these methods will be given in this section.

The accuracy of the calculation of a reserve depends on the data available. Obviously, before any exploration has been initiated these data are quite incomplete, while the maximum amount of data has been gathered at the end of a field's production. Further, the actual reserve depends heavily upon the production techniques that can be applied profitably. A change in production technique during the production history yields a change in the expected-ultimate reserve. Since a company normally is interested only in those reserves that can be produced profitably, a change in economic conditions may change the reserve as well. The expected-ultimate reserve is in fact a continuously changing figure; it varies as more data are gathered, and depends on technical and economic developments.

Four essentially different methods can be applied to estimate an oil or gas reserve:

(1) Estimation by comparison of geological and technical data from other areas or fields.

(2) Volumetric estimation.

(3) Estimation by the material-balances method.

(4) Estimation with production-decline curves.

The above ordering mirrors roughly the chronological development of calculation methods during the lifetime of the field. If not a single exploration well is drilled in the concession area, the reserve to be expected depends on guess-work. This guess-work can be highly professional as will be illustrated in Chapter IV, but it remains guess-work. As soon as the field is discovered and enough production data become available, the production-decline-curve method can be applied. The uncertainty is, of course, much greater with the first method than with the last. The other two methods fall somewhere in between.

Estimation by comparison

This method results in conjectures with a very wide margin of error, usually beginning at zero, because it is uncertain whether producible oil or gas is present at all. Unfortunately, the reserve figures are most urgently needed during the first phase of exploration when large investments for exploration work—especially offshore—must be made. Therefore, a guess with a wide margin of error is better than no guess at all.

The comparison method can be employed only if the exploration concession is situated close enough to existing producing fields—or at least near areas where relevant information has been gathered about the geology and production history. Such information can be plotted on maps, constructed with contour-lines encircling areas with the same amount of oil or gas beneath the surface per surface unit. An example is the "barrels-per-acre"-map. Such maps may give insight into the amounts that might be expected beneath the concession area. Maps ordering exploration results, such as success -ratio maps, may also be constructed (see DOWDS, 1968).

Volumetric estimation

The accuracy of the volumetric estimation depends entirely upon the accuracy of the different parameters. The volumetric calculation can be used even before drilling based on estimations of these parameters. A more accurate volumetric estimation can also be performed following an extensive drilling program. The volumetric calculation is essentially the volume of oil or gas in place multiplied by the recovery factor. The recovery factor is the percentage of the amount of oil or gas in place that can be recovered. This figure depends on the drive mechanism believed to be present and the recovery techniques applied. The parameters to be used for the calculation of the volume of oil are slightly different from those that are necessary to calculate the volume of gas.

Volumetric calculations of recoverable oil. The recoverable oil can be estimated with the following formula:

$$N = A \times h \times \phi_e \times (1 - S_w) \times \frac{1}{B_{oi}} \times E_R$$

The different parameters in this formula have the following meanings:
(1)   The area of closure or the area that is underlain with oil is expressed as A.
(2)   The average thickness of the oil-containing reservoir rock is given by h.
       Instead of multiplying the surface by a thickness, the volume of the reservoir rock can be obtained directly from an isopach map by planimetering. An isopach map contours points of the same net-reservoir-rock thickness in the reservoir rock. The volume can be obtained directly from these maps. In this case $A \times h$ can be replaced by V which is the net-reservoir-rock volume. The isopach map can be obtained from seismic- and well data.
(3)   The effective porosity is given by ϕ_e. It comprises that part of the porosity consisting of interconnected pores, meaning that the oil can flow from these pores to the bore hole. In the formula the weighted average must be used, if possible, of the various measurements in the different wells.
(4)   From the pore-space must be subtracted that part filled with water. Water is generally present in thin films covering the sand grains or other parts of reservoir rock. The water content is expressed in the fraction S_w. The weighted average must also be used in this case.
(5)   The volume of oil must be adjusted to equal itself at surface conditions; this correction is indicated by B_{oi} (see FRICK, 1962).
(6)   The recovery-efficiency factor E_R is the fraction of the oil in place that can be recovered. This fraction is normally very difficult to fix, especially when the drive mechanism is not yet known, or known only sketchily. The drive mechanisms may be, for instance, water drive, gas-cap-expansion drive, or dissolved-gas drive (see ARPS, et. al., 1967).
(7)   Finally, the combination of the previously named parameters leads to the amount of recoverable oil: N.
       The volume of recoverable oil obtained in this way has necessarily a wide margin of error, even when some of the parameters can be estimated fairly accurately.

Volumetric calculation of recoverable gas. The volumetric calculation of recoverable gas depends essentially on the law of Boyle-Gay-Lussac, which states that:

$$p \times V = R \times T$$

for each given amount of gas. In this formula p is the pressure, V is the volume, T is the temperature and R is a constant. This means that when gas is brought to the surface, as the pressure and temperature change, the volume changes as well. The calculation of the recoverable-gas volume at surface conditions ($25° C$, 1 atm) consists of the subtraction of the gas remaining in the reservoir from the gas initially in place. The calculation is, with these two variations, very similar to that for the recoverable oil.

In addition; the recovery-efficiency factor for gas is difficult to estimate, as illustrated by the Slochteren-gas field. If the gas is produced with the normal depletion or gas-expansion drive the proved reserve is estimated at 1900×10^9 m^3. If a considerable water drive takes place the recoverable gas is thought to be much less, only 1450×10^9 m^3 according to WELLS (1968).

The calculations of recoverable gas from associated gas, dissolved gas, or natural-gas liquids is possible with variations of the two described techniques.

Estimation by the material-balance method

The material-balance method can be applied only if a certain production history of the oil or gas field is known. Schilthuis showed in 1936 that this method is well suited to fixing the amount of oil or gas initially in place. The method requires extensive arithmetic and will not be discussed in detail. It involves a close study of all significant factors which change during production, including pressure, gas/oil ratio, water volume, etc. The method is based on two certainties: the produced amount of oil or gas and the fact that the original oil in place must have been a certain (constant) amount. The gradual changes in the parameters give insight into the original reserve and consequently into the remaining reserve.

Estimation with the production-decline curves

The most popular variable that is used for mathematical extrapolation to analyze the ultimate-recoverable oil or gas is the production rate. The production rate per time period gives results that can be easily interpreted. The basis of the reserve estimate is the assumption that the future behavior of a well will be governed by whatever trend or mathematical relationship has been apparent in its past performance. The extrapolation, therefore, has nothing to do with geology or reservoir engineering but is purely empirical.

The production rate is usually very accurately recorded, allowing the past performance of the well to be easily obtained. The point in the future where production must stop is based on economics. This point is generally established when the production rate is such that operation costs do not merit continued production. This production rate is the economic limit of the well. If the production of the well were not influenced by disturbing factors, such as legally-fixed maximum-production rates, repairs, or transport difficulties of the oil or gas, the production rate would follow a comparatively smooth decline curve. This curve can be extrapolated by using some mathematical function mirroring the past performance of the well. In the main, the so-called exponential-decline or constant-percentage-decline curve is used, but other mathematical functions including the hyperbolic-decline curve and the harmonic-decline curve are also valuable.

Clearly the simple mathematical extrapolation per well may be considerably disturbed by the previously mentioned factors; additionally, the simple mathematical relationship often proved lacking. The extrapolation is progressively more accurate the longer the production history (see FRICK, 1962, p. 37-42).

Case histories

Before attention is devoted to the calculation of reserves on a large scale, some results obtained from a single pool's reserve calculation will be considered. The original recoverable-reserve calculations may be too high or too low. According to KEPLINGER (1966) upward adjustments have been frequently the result of more efficient outcomes for the drive mechanisms than expected. Downward adjustments are necessary following unforeseen poor-reservoir communication, erroneous estimates of the volume or the recovery factor, or when fractured reservoir conditions prevent the application of successful secondary-recovery techniques. The causes for adjustments are so myriad that it is no wonder that they are frequently made. Of course, the application of new production techniques and economic developments changes the ultimate outcomes, but these changes have no connection with original poor estimates.

The fractured nature of the reservoir, for instance, caused a considerable difference between the originally-applied recovery factor (which was 23%) in the West Edmond Hunton Pool (Oklahoma, U.S.A.) and the final outcome (that proved to be 17%). On the contrary conservative predictions necessitated in the Redwater Field (Alberta, Canada) an upward adjustment of the recovery efficiency from 50 to 64%.

It is difficult, however, to pinpoint the extent to which adjustments are due to new technical developments and economic developments.

Calculation of the reserve beneath a large area

The calculation of the reserve beneath a large area can be carried out in two different ways. All the reserve calculations from the different pools or concessions can be combined into a single figure—this is the indirect method. It is also possible to calculate the reserve directly from other geological assumptions.

The indirect method

The indirect method is frequently applied and is the base for most of the statistical data that are furnished by a number of countries. In this case the reserve figures per pool, per field, per company, etc., are added to arrive at a reasonable national-reserve figure for the specified mineral. The trouble with such additions is that one is lumping reserve figures having different meanings. One company will arrive at rather conservative reserve figures while another company reaches somewhat speculative conclusions. Although the reserve calculations are based on exact material, there are a number of subjective parameters to be included in the analysis—for instance, the expected-recovery factor or the mathematical relation chosen to extrapolate the production-decline curve. To arrive at a national-reserve figure it is necessary to formulate a definition for the reserve figure desired by the government. This definition may lie between two extremes. The most conservative approach is to define the reserve in such a way that no uncertainty exists concerning the presence of the oil and gas and the technical and economic possibilities for producing the minerals. An inquiry based on this definition will yield the smallest figures possible for the national or regional oil or gas reserves. The opposite extreme is to include all the oil pools in the reserve, irrespective of whether they are economically or technically recoverable. This provides the geological reserve or the resource base. Between these two extremes definitions abound, and petroleum literature is overloaded with oil-reserve classifications. The only valid rule that can be drawn is that reserve figures are of no value unless what is meant by "a reserve" is exactly specified. To give an example of the possible extreme reserve figures, the UNITED STATES DEPARTMENT OF THE INTERIOR (1968) states that the resource of the United States is about 2,000 x 10^9 bbl of crude oil, while the American Petroleum Institute puts the "proved reserve" at 31 x 10^9 bbl (see LOVEJOY and HOMAN, 1965). The "proved-reserve" figures of the A.P.I. are the most conservative ones that can be obtained. The "proved-reserve" concept plays an important role in American energy policy and therefore this concept will be examined more closely. Examples are given for oil, but similar definitions exist for gas and natural-gas liquids.

The proved-reserve concept. A special committee of the Society of Petroleum Engineers in the U.S.A. redefined the original definition of the American Petroleum Institute (A.P.I.) in 1965. The following definition has now been adopted by most of the important petroleum organizations in the U.S.A.:

Proved reserves are "the quantities of crude oil, natural gas and natural-gas liquids which geological and engineering data demonstrate with reasonable certainty to be recoverable in the future from known oil and gas reservoirs under existing economic and operating conditions. They represent strictly technical judgments, and are not knowingly influenced by attitudes of conservatism or optimism".

The words "under existing economic and operating conditions" exclude all those reserves that are possibly economically-recoverable with new techniques already applied in other oil fields. The concept of the proved reserve is therefore a conservative one. This definition has, however, the advantage of supplying each year consistent and reliable information. The conservatism of the concept contributes to a situation in which the increase in "proved reserves" for the U.S.A. is owed mainly to revisions of previous estimates, rather than actual newly discovered reserves. Some of the additions to the proved reserves of the U.S.A. derive from revisions of the reserve figures of actual discoveries from before World War I.

The economically-recoverable reserves. The concept of the proved reserves is conservative and for some decisions in energy policy, figures of a less rigid type are more suitable. The UNITED STATES DEPARTMENT OF THE INTERIOR Report (1968) reveals for instance, that the American Petroleum Institute regards in addition to the proved reserves, an amount of 7.4 x 10^9 barrels of crude as "economically available by application of fluid injection, whether or not such a program is currently installed". Other organizations in the U.S.A., such as the Interstate Oil Compact Commission, are convinced that even higher figures may be given for the reserves that are presently economically recoverable.

In the following thesis the concept of the "economically-recoverable reserves" will be used also, but with slight modifications. The extent of the economically-recoverable reserves is a function of the accompanying economic conditions as interpreted by the companies, including mining legislation. For instance, the "economically-recoverable reserves" will be greater the more favorable the mining legislation is to the companies. For a theoretical study of the influence of mining legislation it is, therefore, interesting to regard the "economically-recoverable reserves" as a function of different hypothetical economical and juridical conditions. The economically-recoverable reserve used in Chapter VI of this book is, therefore, a hypothetical concept.

The direct method

With the direct method the reserves are estimated directly, rather than by merely adding results reported by different companies. Outstanding work has been done in this field by the American petroleum geologist WEEKS (1966). In his approach the total amounts of oil and natural-gas liquids recoverable by conventional methods from the world offshore areas were estimated. The operation was essentially a volumetric calculation on a very large scale. He considered the continental shelf to a waterdepth of 1,000 ft (300 m). The first step was to estimate the area and volume of sediments beneath these shallow waters, area by area. Each region was then classified according to its likelihood of containing petroleum. The final step was to estimate the amount of petroleum economically recoverable by present and probable future techniques and economics. In this way the offshore shelves were projected to contain about 700 x 10^9 bbls of oil and natural-gas liquids. The same technique can be employed, of course, for estimates for natural gas and for onshore areas.

Distributions

The distribution of several parameters related to the oil and gas fields can be influential during the preparation of mining legislation. For political, transport and trade problems the geographical distribution of oil and gas fields is a key factor. For the geologist, the distribution of various geological parameters is of great significance. An interesting study has been done in this field by KNEBEL and RODRIGUEZ-ERASO (1956), who classify a number of geological characteristics, such as the type of trap, the lithology of the reservoir, the depth of occurrence, etc., for all the major oil fields. Distributions frequently used are the size-frequency distribution of the oil and gas reserves and the production per field. These two types will be studied more closely.

Size-frequency distribution of the reserves

Extensive study in size-frequency distribution of the reserves has been done by ARPS (1961), KAUFMAN (1962,1965), and McCROSSAN (1968). From KAUFMAN (1965, p. 109), the following remarks are cited: "Clearly the functional form used to characterize the size distribution of oil and gas fields is a vital part of any model which you as decision makers might use to analyse exploration decisions. Ideally, we would like this form to be flexible enough to fit a wide variety of empirical histograms of oil and gas fields in differing areas with differing definitions of reserves by varying only the value of the parameters of the form, not the form itself. We also would like it to be analytically tractable, so that it may be easily used in the course of a formal analysis of exploration decision problems; e.g. by use of statistical decision models. The lognormal functional form has these properties . . ."

Kaufman concluded that reserve-figures for southern Louisiana and Oklahoma were reasonably in accordance with the lognormal distribution.

A study as proposed by Kaufman runs into a number of troubles. First, all those fields producing less than a 1,000 bbls per day were not included in his figures; second, no information exists concerning submarginal fields. Further, the definition of reserve is vague, leading to figures which cannot reliably be compared. Finally, all the discovered reserves are not "ultimate" figures; the given figures will grow in the future through extensions and revisions of the original estimates.

It is, therefore, questionable whether the lognormal distribution would fit if all the relevant geological facts were known.

McCROSSAN (1968) has found lognormal-size-frequency distributions for the available data from the Canadian oil fields.

Another analysis of the U.S.A. oil fields has been made by ARPS (1961). He obtained lognormal distributions for the size-frequency distributions. The results are shown in Table II. He made the following comment concerning his figures: "You may note that there is a strong similarity between all of them (the four distributions). About 5% of the fields, the best ones, seem to contain 50% of all the reserves. At the same time the last 60% of all the fields (the poor ones) seem to yield 5% or less of all the reserves. This last sub-marginal category is the one we should learn to stay out of. They do enter into statistics as wildcat successes, but they don't make any money. The 35% group in between could be termed "profitable" since it appears to contain 45% of all the reserves. The group one should really be after is the so-called "bonanza" group, the top 5% of which contains 50% of all reserves."

These findings are highly significant, since it can be assumed that other areas of the world, if they are large enough to contain a sizeable number of oil fields, will have similar

TABLE II

DISTRIBUTION OF OIL AND GAS FIELDS IN THE U.S.A.[1]

Percentage of total-proved-ultimate recovery	Percentage of all fields, in descending order containing the given percentage of total-proved-ultimate recovery			
	all U.S.A. oil fields	new U.S.A. oil discoveries	new U.S.A. gas discoveries	Denver-Julesburg-basin
25	0.34	0.9	1.4	1.3
50	1.6	4.2	6.4	5.2
75	6.0	12.4	15.6	15.0
90	20.0	26.0	28.5	29.6
95	33.0	39.0	38.4	40.8

[1] Source: ARPS (1961, p. 159).

distributions. Thus those engaged in the preparation of mining legislation must be aware that the bill's area of jurisdiction will probably contain a few bonanzas and many marginal fields—and the majority of the reserves will be concentrated in these bonanzas.

Size-frequency distribution of the production per field

Reserve-figures have the disadvantage of being difficult to obtain. Production-figures, on the contrary, are easily available. Therefore it is interesting to chek the results for the reserve distributions with production figures. On the average, fields containing large reserves will produce large quantities per year. The relation is not such that the yearly production will be a constant fraction of the reserve of each field, because in the early days of a large field's lifetime production is building and yields are low with regard to the entire reserve. Further, the lifetime of large reserves will be longer on the average than small reserves. In general, however, the same conclusions can be assumed for the production-figures as for the reserve-figures.

Table III shows clearly that the majority of the production in a substantial number of countries is issuing from bonanzas. About 20% of the fields (the best ones) contribute to 70% of the production for 18 major oil-producing regions in the world outside the U.S.A. and the Communist area. The figures are, of course, somewhat incomplete since

TABLE III

CONTRIBUTION OF THE LARGER FIELDS TO THE TOTAL PRODUCTION IN PERCENTAGES FOR THE MOST IMPORTANT OIL PRODUCING COUNTRIES IN THE WORLD OUTSIDE THE U.S.A. AND THE COMMUNIST AREA

Country	Total number of fields	Percentage of the total production contributed by the largest fields	
		number of large fields	percentage
Algeria	30	6	75
Brazil	19	4	73
Canada (Alberta)	60	12	69
Colombia	29	6	66
France (Aquitaine)	9	2	82
Gabon	10	2	67
Indonesia (C. Sumatera)	15	3	88
Iran	22	4	76
Iraq	8	1	69
Lybia	27	5	69
Mexico	52	10	70
Netherlands — NW-Germany	48	10	51
Peru	11	2	62
Saudi Arabia	12	2	73
Trinidad	21	4	71
Turkey	14	3	56
Venezuela (Anzoategui)	40	8	64
(Zulia)	20	4	84

Source: Oil and Gas Journal, 31 dec. 1968.

20% of the mentioned fields does not actually mean 20% of total fields, because the smaller fields are not listed. The general trend, however, of a few fields accounting for the majority of production is attested to by these facts.

From the preceding discussion it can be concluded that the majority of the reserves and the production is concentrated in a few large fields in a selected area. This conclusion will be intensively used in Chapter VI.

Literature

Arps, J. J., 1945. Analysis of decline curves. *Trans. A.I.M.E.*, 160: 228-247.

Arps, J. J., 1956. Estimation of primary oil reserves. *Trans. A.I.M.E.* 207: 182-191.

Arps, J. J., 1961. The profitability of exploratory ventures. In: INTERNATIONAL OIL AND GAS EDUCATIONAL CENTRE SOUTH WESTERN LEGAL FOUNDATION (Editor), *Economics of Petroleum Exploration, Development, and Property Evaluation.* Prentice Hall, Englewood Cliffs, N.J. pp. 153-173.

Arps, J. J., Brons, F., van Everdingen, A. F., Buchwald, R. W., and Smith, A. E., 1967. A statistical study of recovery efficiency. *Am. Petr. Inst.*, D 14: 1-33.

Campbell, J. M., 1959. *Oil Property Valuation.* Prentice Hall, Englewood Cliffs, N. J., 523 pp.

Dowds, J. P., 1968. Mathematical probability approach proves successful success ratio during 1964-'68 averages better than 50% for 118 wells, including wildcats and extensions. *World Oil*, 167(7): 82-85.

Frick's Petroleum Handbook, 1962. McGraw Hill, New York. Chapter 37: Estimation of primary oil and gas reserves.

Hughes, R. V., 1967. *Oil Property Valuation.* Wiley, New York, N.Y., 313 pp.

Kaufman, G. M., 1963. *Statistical Decision and Related Techniques in Oil and Gas Exploration.* Prentice Hall, Englewood Cliffs, N. J., 307 pp.

Kaufman, G. M., 1965. Statistical analysis of the size distribution of oil and gas fields. In: SOCIETY OF PETROLEUM ENGINEERS (Editor), *Symposion on Petroleum Economy and Evaluation.* Soc. Petrol. Engrs., Dallas, Texas, pp. 109-124.

Keplinger, C. H., 1966. Case histories of actual performance of appraisal prognostication-petroleum reservois. SOUTH WESTERN LEGAL FOUNDATION, Editor *Exploration and Economics of the Petroleum Industry*, 5: 281, 307.

Knebel, G. M. and Rodriguez-Eraso, G., 1956. Habitat of some oil. *Bull. Am. Assoc. Petrol. Geologists*, 40(4): 547-561.

Lovejoy, W. F. and Homan, P. T., 1965. *Methods of Estimating Reserves of Crude Oil, Natural Gas and Natural Gas Liquids.* RESOURCES FOR THE FUTURE. Johns Hopkins Univ. Press, Baltimore, Md., 163 pp.

McCrossan, R. G., 1968. An analysis of size frequency distributions of oil and gas reserves of western Canada. In: GEOLOGICAL SURVEY OF CANADA, DEPARTMENT OF ENERGY, MINES AND RESOURCES (Editor), *Report of Activities, Part B, Nov. 1967 — March 1968.* Roger Duhamel, Ottawa, Ont.

United States department of the interior, 1968. *United States Petroleum through 1980.* U.S.Dept. Interior, Washington, D.C., 92 pp.

van der Laan, G., 1968. Physical properties of the reservoir and volume of gas initially in place. In: KONINKLIJK NEDERLANDS GEOLOGISCH EN MIJNBOUWKUNDIG GENOOTSCHAP (Editor), *Symposion on the Groningen Gas Field.* — Verhandel. Koninkl. Ned. Geol. Mijnbouwk. Genoot., Geol. Ser., 25: 25-33.

Weeks, L. G., 1966. Assessment of the world's offshore petroleum resources and exploration review. *Exploration Econ. Petrol. Ind.,* 4: 115-148.

Wells, M. J., 1968. Dutch drilling starts offshore and on land. *World Petrol,* 39(5): 27-29, 72.

Chapter III

Petroleum economics and law

The purpose of this book is to analyze the fundamental framework of petroleum legislation. Before this can be done successfully, some essential economic and juridical problems must be outlined. Each country forms a part of the world economy. This is especially true of the world-petroleum economy, since petroleum products account for nearly 60% of the world's energy consumption (UNITED NATIONS, 1968, p. 15). The economic development in the coming decades will be partially determined by the way the world oil and gas stream can be continued and expanded.

This chapter contains some notes about the world-oil market, about gas markets and some related juridical problems. It must be mentioned that this chapter concerns only the exploration, production, and transport of oil and gas, while refining and marketing activities are excluded from the analysis.

The world-oil market

The geological reserve of oil and natural-gas liquids in the world is considerable. LAMBERT (1966) gives a figure of 11×10^{12} barrels or 1.8×10^{12} m^3.

Only a small portion, about 400×10^9 barrels (60×10^9 m^3), is considered "proven". The majority (about 75%) of the proved oil reserves is located in the Middle East and North Africa.

World oil production amounts to about 15×10^9 barrels per year. The proved reserve in roughly 25-30 times the yearly production. This does not mean, of course, that the world oil reserve will be exhausted within three decades. New oil must and can be found as soon as oil is produced. In the U.S.A. and the U.S.S.R., the proved reserves are maintained at about 12 times the yearly production (ADELMAN, 1965, p. 66). It is not economical to seek more reserves if they are not needed within a foreseeable time span.

The large concentration of oil in the Middle East and North Africa causes typical problems. About 1% of the world population is producing approximately 40% of the world's oil. This area is consequently the main exporter of oil. Consumption of oil is also markedly disproportionate, since the industrialized countries in particular need energy. The U.S.A., western Europe, and Japan consume about 70% of world-oil production, while these areas contain only 20% of the world's population. The main stream of oil is therefore from the oil-exporting and developing countries to the industrial states. More than half of the marine shipments is performed by oil-tanker traffic, to keep this stream flowing (Table IV).

Outside the U.S.A. and the Communist bloc, the world-oil market is controled by

TABLE IV

WORLD CRUDE OIL PRODUCTION AND PROCESSING

(10^6 bbls per day)

Production		Processing	
U.S.A.	10.8	U.S.A.	10.6
Canada	1.3	Canada	1.2
Other Western Hemisphere	5.1	Other Western Hemisphere	4.4
Europe	0.4	Europe	11.2
Africa	5.2	Africa	0.6
Middle East	12.4	Middle East	2.0
Far East and Australia	1.1	Far East and Australia	4.8
Communist Area	7.2	Communist Area	6.4
Total	43.5	Total	41.2

Source: Jaarverslag N.V. Kon. Ned. Petrol. Maatsch., 1969.

seven international, Anglo-Saxon, companies: ESSO, BP, SHELL, GULF, TEXACO, SOCAL, and MOBIL. Their influence is under heavy attack from smaller German, French, Italian and Japanese companies and American independents.

The oil-exporting countries want higher shares in the huge rents that are earned from working the rich oil fields within their borders. Furthermore, the world-oil-political situation is complicated by the growing influence of the U.S.S.R. and the explosive situation stemming from the Arab—Israel conflict.

The investments by the petroleum industry are enormous. It is estimated that the average capital cost to make one barrel per day available for consumption is about U.S.$ 2,700.—. These costs can be (roughly) broken down as U.S.$ 800.— for production and pipelines, U.S.$ 550.— for marine transport, U.S.$ 550.— for manufacturing and U.S.$ 600.— for marketing; while U.S.$ 200.— per daily barrel is needed for exploration (SHELL BRIEFING SERVICE, 1967, p. 3). In 1968 the total capital expenditures, including exploration, in the non-Communist world were U.S.$ 19.2×10^9 (Table V). (ANONYMOUS, 1970a). These outlays are expected to rise 5% annually (HELLER, 1968, p. 42) for the coming decade. More than 90% of these capital expenditures are made by private companies, and the outlays are financed in 75-80% of the cases by self-generated cash.

The world-oil market will be given a more thorough treatment in this section. The opening, general remarks will largely follow the excellent work of ADELMAN (1965). The analysis will concentrate on the increasing or decreasing costs aspect of the oil industry and on the main destabilizers of the world-oil market. The history and the present situation of the world-oil market will also be studied.

TABLE V

NON-COMMUNIST WORLD: OIL INDUSTRY CAPITAL EXPENDITURES 1968
(10^6 U.S. DOLLARS)

	Pro-duction	Trans-porta-tion	Pro-cessing	Marke-ting	Explor-ation	Others	Total
U.S.A.	4,925	475	1,450	1,150	715	350	9,065
Other countries	2,535	2,280	2,980	1,515	615	240	10,165
of which:							
Canada	575	90	195	150	175	15	1,200
Venezuela	185	5	40	5	15	5	255
Other Western							
Hemisphere	400	210	425	100	75	90	1,300
W. Europe	275	180	1,250	825	125	95	2,750
Africa	575	80	65	60	75	5	860
Middle East	300	85	205	25	50	10	675
Far East	225	30	800	350	100	20	1,525
Unallocated*	––	1,600	––	––	––	––	1,600
Total	7,460	2,755	4,430	2,665	1,330	590	19,230

* Foreign - flag tankers.
Source: Petroleum Press Service, Febr. 1970.

Increasing or decreasing cost?

Considerable confusion exists concerning the question whether or not the oil industry is characterized by increasing or decreasing cost. The oil industry is often described as a decreasing-cost industry, which logically leads toward the forming of a "natural monopoly". Large production units produce relatively less expensively than smaller ones, due to the "economies of scale", implying that it pays for larger production units to add incremental production at the cost of the smaller production units, with a "natural monopoly" as the result.

ADELMAN (1965) shows the limits of this theory with regard to the oil industry in two examples. These limits are the size of the market and the factor time. The two examples are pipelines and tankers.

A pipeline company may exemplify a decreasing-cost industry. The larger the diameter of the pipeline, the lower the cost per unit of transported oil. If the market is small, only one pipeline will be necessary to supply the market. If the market is growing beyond a certain volume, more pipelines are needed. An illustration is provided by the north American gas market, which is supplied not by one big pipeline but by numerous smaller ones, due to the large size of the market, and its complicated distribution pattern. For the tanker example, ADELMAN (1965) is cited: "Consider now the world-tanker fleet, which serves a single market (with the partial exception of the U.S. coastal traffic). Taken as a whole, that fleet is one of the clearest examples of increasing cost, because tankers form an array from lowest to highest cost. When, as during the winter of 1962-63, demand is especially high, more and more inefficient tankers are pulled out of layup, or out of the grain trade, and the marginal cost and the market price of the transport service are higher.

In longer perspective, the increasing cost is moderated because many more ships can be built. The pressure is transmitted to the shipbuilding yards, and therefore greater output still means higher cost, though to a milder degree. In the very long run, however, given time enough to adapt shipyards also, the tanker industry is one of approximately-constant marginal and average cost. But there can be a striking-technical progress and cost reduction over time. At any given time, therefore, taking the tanker fleet as it exists and will exist, it is an increasing-cost activity. In 1970, it will still be one, but the whole range of cost, or most of it, will be lower than it is today."

The principal of the increasing marginal cost in the short run and the decreasing cost in the long run will be illustrated with a few examples in the next paragraphs, accompanied by some cautionary critical remarks.

Short term: increasing marginal cost for a single production unit

Exploration is a typical long-term activity. Therefore, it is not reasonable to include exploration outlays in this analysis. If more oil must be produced on a short-term basis, additional production must be met from existing facilities, and marginal cost will be determined by the operating costs. If there is a few month's time lapse, additional production wells can be drilled to increase production in existing fields and marginal cost will be determined by both operating costs and variable capital costs. STEKOLL (1961, p. 145) studied operating costs for wells within the same field. He concluded that these operating costs for an onshore example in Texas varied considerably per well. The cheapest well costs U.S.$ 97.— per month, while the most expensive was U.S.$ 244.— per month. This means that, if the wells produced about the same amount of oil (which is questionable), even among the producing wells in a single field clearly increasing marginal costs exist. For an offshore drilling platform deviations from this rule may occur. CRONEN (1969) calculated that on a platform in the North Sea capable of handling twelve wells, the first ten wells add proportionally to the operating costs, while expenses diminish with the operation of eleven or twelve wells. This is an example of decreasing marginal operating costs per well. The capital outlays for these two extra wells can, however, be higher if their drilling must be highly deviated (and consequently, measured along the bore hole, deeper) than the first ten wells. On the other hand, capital outlays for production platforms, etc., can be distributed over a larger production. The question of increasing marginal cost with offshore conditions depends, therefore, on the actual situation.

*Short term: increasing marginal cost for a **number** of production units*

If the supply curve in the sum of a number of production units (leases, fields, oil-producing basins) is studied, a typical arrangement may appear, as given in Fig. 3. A similar graph was drawn by BRADLEY (1966, p. 2) for different oil-producing basins but it is equally applicable to other production units. Assume that a region contains a rich field, some fields of medium quality and a number of marginal fields. The supply curve for the rich field will show low marginal costs for most of the wells, and rising marginal costs for a few additional wells, until a new field can produce for the same costs and so

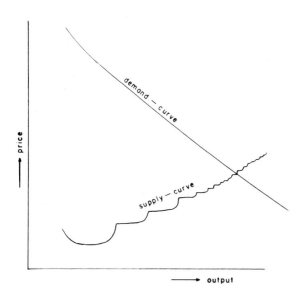

Fig. 3. The supply curve for a number of oil-producing fields

on (rich fields may, of course, also contain marginal wells but these are omitted for simplicity). The price will be determined by the marginal fields. A number of reasons contribute to the assumption of large cost differences between fields.

Production. An important reason for "cost jumps" in the supply curve is because oil fields occur at different depths. Drilling costs for a deep oil field are higher than for a shallow one. FISHER (1964, p. 42) gives a mathematical relation for the drilling costs with depth at a given point in time: $Y = K \times (e^{aX} - 1)$

In this formula Y represents the drilling costs per well, X the depth; e is the base of the natural system of logarithms, while a and K are constants. Consequently, if two fields are drilled at the same time, the deeper field may cost progressively more. But drilling costs may also differ considerably when fields are found at roughly the same depth, as can be seen from the study of HOUSSIERE and JESSEN (1969, p. 55), where the average drilling costs per foot of a number of deep fields (below 15,000 ft) in the U.S.A. are compared. In this case drilling costs vary from U.S.$ 28.— per foot to U.S.$ 88.— per foot (The deepest producing gas field in the U.S.A. lies at 22,800 ft (6,800 m)). Offshore, "economies of scale" may play an especially important role. CAMERON (1966) compares platform costs for use in 200 ft of water. A 1-well, 4-pile platform is estimated at U.S.$ 350,000.— while a 24-well, 16-pile self-contained platform would cost U.S.$ 1,750,000.—, or U.S.$ 73,000.— per well!

In addition to production, similar supply conditions exist with exploration and transport.

Exploration. FISHER (1964, p. 3) constructed a supply curve of exploratory effort by comparing the different oil-producing regions in the U.S.A. Although his supply curve bears no relation to that designed in Fig. 3, his conclusions provide some interesting information. He found that an increasing marginal cost characterizes exploratory effort, and that a substantive distinction can be made between the supply curve of exploratory effort and the supply curve of new petroleum discoveries. For instance, a price increase of 1% for oil at the wellhead resulted in an increased wildcat drilling of 2.85%, but the success ratio *declined* by 0.36%, and there was a decline in the average size of the discovery by about 2.18%. Consequently, the same price increase leads to an increase of new discoveries of only 0.31%. The given percentages can be regarded only as an order of magnitude, but Fisher arrived at an interesting conclusion. A price increase leads to greater exploratory activity, but the search is concentrated upon more risky and smaller projects.

Transport. Transport supply curves have again nothing to do with those seen in Fig. 3, but for the industry as a whole these curves are also important. The essentials were given previously in the quotation from ADELMAN (1965), but two examples will be given in addition.

HUBBARD (1967) figures that the costs for pipeline transports are U.S.$ 3.00/1,000.— bbl miles for a yearly transport of 10,000 bbls, while these costs decrease to 12.5 U.S.$cents/1,000 bbl miles when 10^6 bbls per year are transported. A classic example for the steeply-rising, short-term marginal transport costs was provided during the Suez-crisis of 1967 when 40% more tanker capacity was needed, causing a rise in tanker-charter-freight rates of around 400%.

Rent and quasi rent. Following a discussion of some aspects of the increasing marginal costs for the production of oil in the short term, Fig. 3 merits another look. The marginal-cost curve for an individual field is illustrated by the rich field. Marginal costs are declining for relatively small outputs from the field due to the diminishing impact of the capital charges for equipment and pipelines. When a certain output is reached, capital charges and operating costs are nearly proportional to the expanding output; finally, increasing marginal costs make an appearance as less productive areas in the field must be brought in production and when the field as a whole is reaching its maximum capacity. New increments in production must come from new fields; but since the marginal-production costs of these fields are higher, new fields can best be kept in production until the marginal costs equal the revenues. The reasons for the differences in marginal costs are for a large part determined by the natural occurence of the oil. It is assumed that the marginal costs include a "normal profit", being, according to STONIER and HAGUE (1964, p. 127), that profit which is just sufficient to induce the entrepreneur to stay in the industry. For an oil company this means that it can continue exploration for replacements of oil produced from the already discovered fields. If this normal profit is included in the costs, a considerable rent, or surplus profit, is earned on the rich fields. If the complete characteristics of oil fields were known even before discovery, this rent would be earned by the government or the landlord (under conditions of perfect competition) because any company would be ready to pay this rent to obtain the acreage. The profitability or unprofitability of a field is unfortunately not known in advance. Consequently, it is impossible to pay all the rents to the government before production is begun. Companies are unwilling to consign large sums for the right to explore or produce a concession because the element of risk is large. If a rich field is discovered, however, the government is deprived of a large rent. In both cases one of the parties—the company or the government (landlord)—is unlucky. Mining legislation must therefore be able to cope with all possible outcomes of exploration.

A few words remain to be said about the "quasi rent", which has much to do with the definition of "normal profit". In the short run it seems that the company is earning excessive profits from production of a rich field. It must, however, be remembered that the discovery of such a "bonanza" is a rare occurrence. If the company wants to stay in the industry it must go on exploring for more oil; yet, during this exploration a large number of dry holes will be drilled. These dry holes are an intrinsic feature of the industry. Normal profit, in such a situation, must be such that these failures can be met. (It is rather difficult, however, to distinguish between failures due to the random element in exploration and the failures due to bad management. The latter failures, of course, must be excluded from the "normal profit" determination.) For a government—especially when other exploration activities occur outside the border of its jurisdiction—these dry holes are not relevant for inclusion in the "normal profit", and only exploration costs directly attributable to the field are considered. Consequently, there is a profit that is regarded as a true rent by most governments—which, in the short run, companies can do without, but in the long run need to survive competitively in the industry. Therefore a part of the "rent" must be regarded as a "quasi rent", because it is a surplus in the short run but not in the long run (ADELMAN, 1965, p. 76). This leads the discussion toward other long-term problems.

Long term: decreasing costs

In the long term, continuous technical development can be expected to produce decreasing costs for oil *production*. A rich field producing at present will probably be an even richer field within a decade (if production can be maintained) due to technological progress. This progress is proceeding rapidly. FISHER (1964, p. 43) concluded that, although drilling costs were higher at greater depth, and although the average well depth was growing during the years 1956-1959, drilling costs actually decreased during the same time period due to technical progress.

The NATIONAL PETROLEUM COUNCIL (1967, p. 15) reported that the total production-cost savings per barrel for the period 1950-1965 due to technological and management advances were about U.S.$ 1.00 per barrel in the U.S.A. Four factors played an important role: decreasing drilling costs, decreasing operation costs, better corrosion control, and the trend toward using a wider well spacing (permitting the production of roughly the same amount of oil with fewer wells).

On the other hand, the picture for *exploration* is less clear. HODGES and STEELE (1959, p. 88-97) stressed that the major peaks in the discoveries of new reserves in the U.S.A. can be correlated with the introduction of new exploratory methods. These three peaks occurred in 1911, 1928, and 1938. The first era of exploration can be characterized as one of surface geology. The discovery of the tremendous oil reservoir of Spindletop, near Beaumont, Texas, in 1901 initiated this period of exploration which reached its peak around 1911. Around 1924 a second era began—one employing extensive geophysical methods such as the magnetometer and the torsion balance, and which was supplemented by subsurface-logging methods. This resulted in the peak of 1928. Finally the era of the intensive geophysical methods followed with its seismic exploration techniques. The peak

TABLE VI

OFFSHORE COST FIGURES

Significant cost figures are included in a recent Royal Dutch/Shell study of exploration and production operations in deeper waters. Offshore expenditure for existing depth ranges is already some three to five times that on land and costs—particularly development costs—rise rapidly as the water depth increases. Taking the North Sea as an example: a fixed platform in 100 ft. of water costs about U.S.$ 3,500,000.—; in 200 ft. of water it would cost an estimated U.S.$ 4,750,000.—; in 300 ft. more than U.S.$ 7,000,000.— in 400 ft. about U.S.$ 10,500,000.—; and in 500 ft. U.S.$ 14,250,000.—.

Distance from shore is also a vital consideration if a pipeline is part of the development plan. In the North Sea the 35-mile 30 inch natural-gas pipeline from the Leman field to the shore costs nearly U.S.$ 20,000,000.— almost U.S.$ 600,000.— a mile. The cost here was greatly influenced by hold-ups due to bad weather, and the weather factor must be taken into account when estimating for any given area.

TABLE A: COMPARISON OF OFFSHORE AND ONSHORE DEVELOPMENT OF AN OIL FIELD
(U.S. $ 1,000.—)

	Offshore	Onshore
platforms	1,596	———
production facilities	1,176	1,064
flowlines	2,699	672
flare gas facilities	644	70
well sites	3,696	56
trunkline to terminal (16 km 14 inch)	1,044	756
communication and control	181	140
marine transport and equipment	1,965	815
road transport equipment	269	420
road system	280	742
COMPARABLE DEVELOPMENT COSTS	13,510	4,735
well costs	11,480	3,920
TOTAL	24,990	8,655
terminal and office facilities	13,762	13,762
GRAND TOTAL	38,752	22,417

Estimates based on 14 wells — 12 well jackets — 120,000 bbl/day production rate, 16 km from terminal.

TABLE VI (continued)

TABLE B: COMPARISON OF OFFSHORE AND ONSHORE DEVELOPMENT OF A GAS FIELD
(U.S. $ 1,000.—)

	Offshore	Onshore
platforms	14,000	——
production facilities and foundations	6,160	4,480
trunkline (50 km 30 inch)	21,280	——
gathering line (2 km 30 inch)	1,400	420
treating plant	6,300	6,300
offices and buildings	280	280
roads	280	952
communication and control	700	420
coastal protection-site improvement	980	280
SUBTOTAL	51,380	13,132
cost of wells	18,000	5,320
GRAND TOTAL	69,380	18,452

Estimates based on: reserve 9×10^{12} cuft.; production 500×10^6 cuft./day; maximum capacity 800×10^6 cuft./day; load factor 60%; two clusters of ten wells.

TABLE VI (continued)

TABLE C: *ESTIMATED YEARLY VENTURE COSTS, ONE OFFSHORE-EXPLORATION-STRING*
           OPERATION (NOT INCLUDING CORPORATE OVERHEADS) (U.S. $ 1,000.−)

	cost range	
payments to contractor	3,300 −	4,500
fuel, lubricants	100 −	200
bits, mud, consumables	450 −	1,000
casing	650 −	800
evaluation	150 −	500
transport	600 −	1,150
shore basis	200 −	250
insurance	300 −	1,150
diving	150 −	350
drilling department	100 −	100
TOTAL	6,000 −	10,000

Cost ranging from relatively easily accessible location in shallow water and moderate environment conditions, to deep water far from shore − rough area.

Source: Petroleum Press Service, Jan. 1969, p. 27.

of these discoveries was reached in 1938. These three phases appear to roughly parallel the discoveries of the giant fields, as is illustrated by HALBOUTY (1968, p. 1117).

Since World War II, the exploratory effort in the U.S.A. (excluding Alaska) has not seemed to offer any spectacular new results. It might be that the absence of revolutionary new developments (D'ARNAUD GERKENS, 1962) in the exploration methods is one of the chief causes. Thus the finding costs per barrel may necessarily be increasing in the U.S.A. now and in the future, although new refinements of the existing exploration methods may lead towards considerable economies.

Another result of the lack of essential new exploratory techniques may be the rising interest in the application of "old techniques" in new areas: the continental shelf and geographically remote areas in Alaska.

Offshore developments are increasing at a high rate. According to GIBSON (1969), offshore production was in 1954 limited to the Gulf of Mexico, Lake Maracaibo and the Caspian Sea, but now drilling operations are carried out in more than forty countries. Outside the U.S.A. and Maracaibo nearly one hundred offshore-drilling units are working, thirty of which are in European waters. These impressive developments were due mainly to two factors: the relatively cheap possibilities for seismic exploration, and the new developments in offshore engineering in building drilling rigs and production facilities.

COOPER and GASKELL (1966, p. 43) mention the relatively inexpensive seismic work offshore. Since it is possible to work with a ship and no shotholes need to be drilled, 200 shots per day can be made at sea against 5 to 10 on land. CAMERON (1966) calculates that costs for shooting the entire North Sea area (5 x 5 mile grid) was in the order of U.S.$ 22.− x 10^6, while shooting a land area of the same size would have cost U.S.$ 88.− x 10^6 (± 50%) or four times as much. This cheap seismic work facilitates the large-scale exploration for possible structural traps. Great advances have been made in new refinements of geophysical techniques, and the computer is giving valuable assistance in handling records—all leading to higher offshore-wildcat-success ratios. Since only large structural traps need to be tested it pays to take the risk of expensive offshore drilling (WEEKS, 1968, p. 41). The costs of continuous offshore exploration lasting one year are estimated to range from U.S.\$ 6.— x 10^6 in relatively easily accessible shallow water to U.S.\$ 10.— x 10^6 for deeper, rougher water. For this price, ten wells can be drilled if conditions are favorable, but very unfavorable situations may permit only a single well to be handled.

The main difference between offshore and onshore costs can be attributed to development outlays (Table VI) that are about three times higher for an offshore oil field and four times higher for an offshore gas field (ANONYMOUS, 1969 a, p. 27).

It can be expected that developments offshore will increase rapidly in the near future. New types of semi-submersible drilling rigs will make possible drilling in deeper water to and beyond the continental margin (at a waterdepth of 600 ft, 200 m). Selfpropelled drillships may decrease the time to move from one location to another. Nuclear-powered, acoustically-controlled sub-sea wellheads will decrease development costs (ANONYMOUS, 1968a, p. 90).

Exploratory activities are also directed toward the less accessible areas of the world. In the U.S.S.R., exploration is being actively carried on in the northern parts of Siberia. In the U.S.A., the recent developments in Alaska, resulting in the discovery of one of the largest petroleum accumulations in the world, are already showing their impact on the world-oil economy. The costs in Alaska are high—up to U.S.\$ 1.5 x 10^6 for each well drilled—or about five times the average for other wells in the U.S.A.

For long term studies of costs it is therefore reasonable to assume that a large portion of the potential decrease of costs due to technical development will be used for drilling and production in areas that were formerly considered to be uneconomical to explore. Unless a "breakthrough" in exploration technology occurs it is therefore questionable whether the supply curve will move continuously towards lower costs in the very long run, when the oil industry is studied as a whole.

Destabilizers

ADELMAN (1965) named two strong destabilizers in the world-oil market: the random element of the finding costs and the inventory aspect of petroleum-production costs.

Unpredictable finding costs

Chapter II showed that the size-frequency distribution of oil and gas reserves may show a lognormal distribution (KAUFMAN, 1963)—or in any case an arrangement such that only a few large fields occur as opposed to numerous smaller ones. It is therefore to be expected that the discovery of a bonanza is a rare occurrence. ARPS (1961) states that the wildcat success rate is about one producer to ten dry holes in the U.S.A.; yet of these producers, roughly 60% must be regarded as marginal fields (see Table II) and only a few large discoveries occur. One giant field is found on an average in 180 wildcat wells. The discovery of a bonanza is to a large extent a question of simple luck. According to HALBOUTY (1968) about 21 giant oil fields have been discovered in the U.S.A. simply by random drilling, and these discoveries may consequently be totally ascribed to luck (the last one occurred in 1950).

Therefore, if a petroleum producer foresees a continuous increase of demand in the long run and decides to initiate exploration to find new reserves, he is unable to know how much this exploration will add to the existing reserve. He may be discovering merely marginal fields, but he can also strike one or two bonanzas. This is an important reason for instability in the oil market. This instability was described by LOVERING (1943) as the under- or over-supply hazard.

The oil fields of the Middle East together form a major disturbance of the distribution pattern of oil fields in the world. In this area have been discovered about 60% of the present world reserves of oil. Although the Masjid-i-Sulaiman field in Iran was found in 1908, it took about twenty years before the main stream of important discoveries began. In the ten years before World War II fields mainly in Iran and Iraq were unearthed, and in the ten years following the war the large discoveries in Saudi Arabia and the small Persian Gulf states were made. In roughly thirty years the world oil reserve increased enormously. This major disturbance of the general oil-market situation is still, and will be for the next decade, an important source of disequilibrium.

Inventory aspect of petroleum-production costs

Dry-hole costs may or may not be attributed to the discovery of a certain field. It is usually very difficult to calculate pastfinding costs applicable to a particular field. ADELMAN (1965, p. 68) says concerning this problem: "Furthermore, even if the finding cost were precisely known, it would be of no interest because it would be a sunk cost, and whether vastly above or below price would not matter . . . What does exist in looking ahead, and does control the price, is the reproduction cost. How much must one expect to spend, taking in the good discoveries with the bad, to reproduce a given reserve?"

The concept of the replacement cost is well known, but the amount is difficult to calculate. LOVEJOY, et al. (1963, p. 15) use it in the following way:

(1) Current-production costs divided by the number of barrels produced; plus
(2) Current-development costs divided by the number of barrels added to proved reserves by development activity ("extensions" and "revisions"); plus
(3) Current-finding costs divided by the number of barrels added to proved reserves by exploratory activity ("discoveries"); are
(4) Adjusted for costs allocated to natural gas and natural-gas liquids.

In this concept three sets of costs related to three different amounts of oil are added into a single figure. The "replacement cost" is consequently not comparable to the economic-cost concept. It is a value that illustrates for the company the effort that must be spent to add in the present a barrel of oil to the known recoverable reserve. Since the reserve concept is a vague concept in itself (cf. Chapter II), it should be kept in mind that the "replacement cost" is vague as well. Yet it does have a definite significance since, if a barrel of oil cannot be sold for a price that is sufficient to meet the replacement costs, the company is not making any progress (even if finding costs attributable to the sold barrel indicate that a profit has been made). The company is then merely replacing expensive (higher than the market price) barrels for cheap barrels in the "inventory". The actual decision to continue exploration is complicated; this decision-making process will be discussed extensively in Chapter IV.

ADELMAN (1965) stresses the destabilizing role of the "inventory". The rate of discount—determined by a number of factors, including types such as political instability—determines whether a company wishes to sell oil from its reserves as soon as possible or whether it prefers to wait. The expected price is important. If it is anticipated that the prices will fall, it is preferable to sell immediately out of stock rather than later. These two factors, the expected rate-of-return and expected price, which are mutually dependent, are disrupting factors in the market. Apart from this, the size of the discovered fields plays a role. It is useless to hold reserves because it costs money.

The inventory aspect of production costs has at the same time a stabilizing factor—following from the fact that it is unnecessary to develop an entire field immediately after discovery. If the market is too small to warrant bringing the entire field into production, it can be partially developed—as is presently the case for a number of fields in the Middle East.

Guided by the work of ADELMAN (1965), a quick overview of some basic elements of the oil market has been presented in these pages. With the theory behind us, we can now proceed to a description of actual developments in the world-oil market.

History of the world-oil market

History until World War II

Until World War II the U.S.A. decidedly dominated the world-oil market. Since Colonel Drake spudded the first oil well in 1859 in Pennsylvania, the U.S.A. produced almost all of the world's oil until 1880 when the Russian production around the Caspian Sea gained importance. In a short four-year period (around 1900) Russia was turning out more than half of the world production (TIRATSOO, 1951, p. 14). However, the U.S.A. quickly regained ascendancy with the tremendous development of the oil industry in California and Oklahoma. Since the U.S.A. was the largest exporter of crude, the world-oil prices were established by the American exporters, and the Russian and Rumanian producers based their prices on the American market. Local prices were sometimes lower than U.S.A. prices, but never exceeded these levels because crude could always be easily bought from American exporters.

During World War I the influence of the U.S.S.R. diminished considerably. Mexico replaced her as the second-largest producer. Mexican production was mainly built up by an individual American miner, Mr. Doheny. He struck the fabulous Casiano No. 7 well which produced in nine years 85×10^6 barrels of crude. World War I helped him considerably with his expansion in Mexico. FANNING (1954, p. 26) writes: ". . . Doheny carried on a one-man-supersales campaign. Almost singlehanded he created the Fuel-oil Age, though World War I gave him a powerful boost. Dohensy's commodity—Mexican oil—gave the Allied navies, merchant marines and industrial plants a superiority which speeded the victory." The "Mexican wonder", however, was over when political and social difficulties forced the American investments out of the country. Since the government's agency,

Petróleos Mexicanos (PEMEX), expropriated the foreign oil companies in 1938, oil production has begun once more to increase slowly.

After the blow-out at La Rosa (1922), attention was drawn toward Venezuela as a potential producer of oil, and this country, under the dictator Gomez, replaced Mexico in the middle American scene (MARTINEZ, 1966). It was primarily Standard Oil of New Jersey (its subsidiary CREOLE) that boosted the Venuzuelan production, together with Royal Dutch/SHELL, and GULF (MENE GRANDE). After 1930, the U.S.A. and Venezuela together dominated the world-oil market and the system of Gulf-plus prices remained the price-setting system in the world. Prices f.o.b. in the Persian Gulf, for instance, were based on the f.o.b. Gulf prices, and discrimination occurred according to the destination of the crude.

The production of Middle East and Far East oil before World War II played a minor role in the world. In July 1928 the famous Red Line Agreement was signed, forming the basis of further developments in the Middle East. The agreement provided for a share in the Middle East production (with the exeption of Iran and Kuwait) for Anglo-Iranian (later BP) of 23.75 %; for Compagnie Française des Pétroles of 23.75%; for Royal Dutch/SHELL of 23.75%; and for ESSO (New Jersey) and MOBIL also of 23.75%; while the wellknown millionaire Gulbenkian obtained 5%. Within the area of the Red Line Agreement the companies agreed not to compete against each other for concessions, or to hold any individual concession without first seeking the permission of their partners (TUGENDHAT, 1968, p. 84). Notwithstanding this agreement, other American companies invaded the Middle East. GULF and the Anglo-Iranian Company founded the Kuwait Oil Company in 1933, and SOCAL (Standard Oil of California) gained influence in Saudi Arabia while attempting to negotiate with the parties of the Red Line Agreement.

World War II again proved the importance of crude oil. Germany and Japan were importers of crude. In the last years before the war the Reich imported 41×10^6 barrels of oil per year, mainly from Standard Oil of New Jersey, SHELL and the Anglo-Iranian. These imports were used to supply a large stockpile that amounted at the beginning of the war to 50×10^6 barrels of oil. The Blitzkrieg tactic was partially developed to meet problems deriving from the German oil shortage—long military activities were for them impossible because oil use had to be confined to the minimum. Since Great Britain did not fall into German hands and the blockade could not be lifted, the oil fields of the Caucasus became of tremendous strategical importance. Thus the invasion of the U.S.S.R. was partially due to the Germans' need to obtain its oil fields. For similar reasons the opening attack by Japan was directed mainly towards the oil fields of the former Dutch East Indies (TUGENDHAT, 1968, Chapter 12).

History after World War II

Production capacity was enlarged considerably in the Middle East after World War II. Oil production in 1946 amounted to nearly 10% of the world-oil production. During 1947 the Red Line Agreement was dissolved and a group of American companies concentrated in the ARAMCO (SOCAL: 30%, TEXACO: 30%, ESSO: 30% and MOBIL: 10%) obtained an enormous concession in Saudi Arabia. This company determined the influence of the U.S.A. in the Middle East, since the influence of the original Red Line

Agreement partners was restricted to Iraq (they formed the Iraq Petroleum Company) and a few Persian Gulf states. The concession area of the ARAMCO was about 300,000 square miles (793,600 km^2) while the I.P.C. obtained nearly 200,000 square miles (450,000 km^2). The dates of expiration of these contracts are 1993 and 2000 respectively. Shortly after the war, prices calculated in the Persian Gulf became aligned with the Gulf prices. As a result the point of equalization (equal c.i.f. prices) moved slowly from the Mediterranean toward Great Britain, as prices increased in the U.S.A. Under the influence of the Economic Co-operation Administration, prices in the Middle East were reduced, and the east coast of the U.S.A. became the equalization point. In 1950 MOBIL began to "post" prices in the Persian Gulf. These posted prices were publicly announced rates applicable to all customers, and were connected indirectly with the Gulf prices by the difference in transport costs to the U.S.A. Within a short time all the major oil companies began similar posting of prices. According to HARTSHORN (1962, p. 138) this posting of prices coincided with a shift in financial arrangements between the Middle East governments and the major oil companies towards the principle of a 50/50 profit split. Before this time, governments of the oil-exporting countries were happy with a simple royalty, but after the war political pressure built up to increase the revenues from the oil production. The result in the majority of countries was a corporate-income tax of 50% of reported profits, based on transfer prices. Due to activities of the Iraq government, however, the posted prices became popular for use in the calculation of the income tax.

The financial discussions between governments and major oil companies led to a conflict in Iran. Under the influence of Mossadegh, an important and nationalistic politician, the Shah was forced to sign on May 1, 1951, a Bill for the nationalization of the Anglo-Iranian Oil Company. About 75% of this company's production was coming from Iran, and the Iranian government gambled on the assumption that it was impossible for Great Britain to maintain her oil imports. The British government assisted the Anglo-Iranian Oil Company, however, in arranging an effective boycott against exports of Iranian oil, resulting in a near stoppage of oil production in that country. In the meantime the Anglo-Iranian Company developed its interests in Kuwait, Iraq and Qatar. These actions promoted a rather rapid recovery in the flow of oil supplies toward Europe. The Iranian government, however, was deprived of nearly all its income until 1954 when an agreement was reached. The nationalization was accepted and the National Iranian Oil Company was formed. A consortium of all the major oil companies operated under this company's name, and the profit-sharing principle of 50/50 was accepted by Iran. The Consortium consisted of Anglo-Iranian (now renamed: British Petroleum): 40%; Royal Dutch/SHELL: 14%; American companies together: 40%, and finally the Compagnie Française des Pétroles: 6%. This agreement again enlarged the American influence in the Middle East considerably.

A few conclusions were drawn from this conflict. The Shah commented: "Mossadegh's big miscalculation lay in his stubborn insistence that he knew how to market our oil with no help from foreigners. Yet at that time we possessed not a single tanker, nor did we have even the beginnings of an international marketing organization" (TUGENDHAT, 1968, p. 145). The companies decided to spread their activities as widely as possible over the world to avoid the risks of similar political troubles in the future.

The supply of crude from the Middle East after this conflict increased markedly, stimulated by price rises in the U.S.A., which were closely followed by the posted prices in the Middle East. The rapid developments in the Middle East resulted in a severe pressure on U.S.A. markets. The Suez-crisis of 1956, however, released this pressure temporarily and no protective measures were taken by the American government until 1958. In the years after the Suez-crisis, tanker rates collapsed and unsettled further the balance of the Middle East crude prices in the world markets (DE CHAZEAU and KAHN, 1959, p. 214). The U.S.A. government adopted in 1958 a system of import restrictions to about 12% of total U.S.A. consumption. The reason for the tremendous influence of the Middle East oil on the American oil market can be studied in Fig. 4. Slight price increases in the U.S.A. brought about a large additional supply from the Middle East, due to the difference in supply curves between the low-cost Middle East oil and the high marginal costs in the U.S.A.

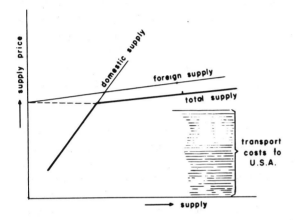

Fig. 4. The influence of Middle East oil on the markets of the U.S.A.

The introduction of the import restriction broke the last weak link between Gulf prices and Middle East prices. These latter prices continued to decrease due to the structural surplus conditions introduced by the import restrictions and because of the unlimited availability of low-cost oil in the Middle East (now made easily obtainable by the decreased tanker rates in the European markets). The destabilizers did their work. This was the beginning of a new era in the world-oil market.

Recent developments

The deterioration of the prices in the world-oil market, and the accompanying decrease of product prices, diminished considerably the profitability of the investments of the major oil companies. The return on invested capital in foreign investments of the oil companies reported by the Chase Manhattan Bank fell from a level of 25% to 30% before the first Suez-crisis to a range of 12% to 15% since 1958 (HILL and COQUERON,

1966). The absolute amount of the capital streams, however, was enormous. SYMONDS (1966) observes: "The United States has been investing for many years past, and on an ever-growing scale, in petroleum abroad. From U.S.$ 248 x 10^6 in 1950, the annual outflow rose to U.S.$ 1,013 x 10^6 in 1965. But the return flow of income has risen even higher—from U.S.$ 555 x 10^6 in 1950 to U.S.$ 1,798 x 10^6 in 1965. As a result, the net sum now coming back on the oil-industry's capital and income account alone has built up to more than U.S.$ 750 x 10^6 a year." (It must be considered that these amounts include transport, refining and marketing activities.)

Organization of Petroleum-Exporting Countries

The deterioration of the world-oil prices outside the U.S.A. had a serious impact on the position of the oil-exporting and developing countries. The posted prices for Arabian crude decreased from U.S.$ 2.12 in June 1957 to U.S.$ 1.84 per barrel in August 1960; while Iranian crude deteriorated from U.S.$ 2.04 to U.S.$ 1.78 per barrel in the same time period. Since the 50/50 profit split was established on the posted prices, the previously delineated price-drops negatively influenced the income of Middle East governments. The same was true of Venezuela where prices decreased from U.S.$ 2.80 to U.S.$ 2.55 per barrel. The main oil-exporting countries decided to defend themselves against this negative development. They founded the Organization of the Petroleum-Exporting Countries (O.P.E.C.) in September 1960. The founding members of this organization were Venezuela, Kuwait, Saudi Arabia, Iran and Iraq. Their principal aim was to co-ordinate the petroleum policies of the Member Countries and to determine the best means of safeguarding their interests, individually and collectively.

The main target was the stabilization of the world-oil prices. The governments wanted the companies to negotiate with them before any change in posted prices was officially established. This resulted in a remarkable price stabilization. The posted prices in the Persian Gulf and in the Gulf of Maracaibo have not changed since 1960. The posted price, however, has become more a "tax-reference price" than a real market price. The major oil companies have been forced to allow important discounts to the buyers of crude. This implies that the realized prices or the real world-market prices are considerable lower in most cases than the posted prices. The revenues of the governments, however, were not directly affected by the further deterioration of the world-crude prices (MARTINEZ, 1966).

For the stabilization of the real world-oil prices since 1960 a rigorous production-restriction scheme would have been necessary. Even if the O.P.E.C.-member countries could have executed this difficult task, it would hardly have been successful. The major oil-companies' policy of spreading their risk resulted in intensified exploration in countries outside the O.P.E.C. A spectacular rise of the oil production in Lybia, Nigeria, Abu Dhabi, Oman, etc. was the result. The discovery of huge new reserves, especially in Lybia, has produced new pressures on prices.

The O.P.E.C.-countries' aims were stabilization of prices and the assurance of a large share of the economic rent. It must be admitted that these two aims of the O.P.E.C. have already been achieved to an important degree. The posted prices have remained stable since 1960, although the realized prices decreased continuously since that time. Presently, Lybia and Algeria are pressing for higher posted prices, because the closure of the Suez-canal makes their oil highly profitable due to relatively low transport costs. The supply of Middle East crude was (and is) so abundant that the companies need occasionally to grant discounts of 25% on the posted prices to sell their oil. The posted prices, however, remained the base of the income-tax calculations.

The share in the rent became continuously larger. The oil revenues for some of the O.P.E.C.-member countries in 1967 were: Iran, U.S.$ 777.$- \times 10^6$; Iraq, U.S.$ 340.$- $\times$ 10^6; Kuwait, U.S.$ 680.$- $\times 10^6$; Lybia, U.S.$ 476.$- $\times 10^6$; Saudi Arabia, U.S.$ 886.$- $\times$ 10^6; and in 1968 for Venezuela U.S.$ 1,248.$- $\times 10^6$ (WELLS, 1968). Although these seem to be large national revenues, on a per barrel basis they are more modest—about U.S.$ 0.85 per barrel for the Middle East and U.S.$ 0.95 per barrel for Venezuela. It is estimated that the capital outflow from the O.P.E.C.-countries to the O.E.C.D.-countries as profits on production and tanker-transport amounts to nearly U.S.$ 4.$- $\times 10^9$ (ANONYMOUS, 1969b).

The success of the O.P.E.C. attracted many new oil-exporting countries such as Lybia, Indonesia, Abu Dhabi, Qatar and Algeria. Consequently, the O.P.E.C. presently accounts for about 72% of the world-oil reserves, 45% of the production, and 86% of oil exports. Their share of refining capacity, however, is only 8%. This last figure illustrates the weakness of the organization: its lack of influence in "downstream" operations, such as transport, refining and marketing (HILL and EALES, 1970, p. 22).

American Independents

Another new development in the world-oil market is the increasing influence of American Independents. The first sign of their activity was in 1954 when a group of independent American oil companies obtained a 5% share in the N.I.O.C. concession in Iran while the five major American oil companies obtained 7% each, making the total share of American oil companies 40%. The big chance for the independents, however, followed the developments in Lybia. The government of Lybia was reluctant to deal with only a few major oil companies and granted a large number of small concessions to as many companies as possible. In 1959, OASIS Oil Company (comprising CONTINENTAL, MARATHON, and AMERADA) found large oil reserves. The motives for these independents going abroad were explained by McLEAN (1968). He stressed four reasons: "First, we believe it is essential from a long-run strategic standpoint for us to have a significant stake in crude-oil reserves overseas. . . . Second, we believe that some day the U.S.A. will need to import larger quantities of foreign crude oil to meet domestic market requirements on an economical basis. . . . Third, we believe that there are still many large, relatively unexplored sedimentary basins abroad that offer attractive investment opportunities. . . . Finally, we believe that the rapidly growing and potentially large petroleum markets abroad will ultimately provide good opportunities for downstream investments."

The aggressive tactics of the "newcomers" have resulted in the fact that about 10% of the western European oil is now supplied by these companies. At the same time the increased competition undercut the position of the "seven majors" and their "oligopolistic"-market structure. This new element must also have been a factor in the decreasing prices for crude oil.

Western Europe and Japan

Another factor of importance in the new world-oil scene are the new initiatives of mainly French, Italian, Japanese and German companies and their governments to gain influence in the Middle East and other prolific oil-producing areas. Western Europe and Japan are both the main importers of crude. Western Europe is importing about 500 x 10^6 tons of crude oil, while the Japanese imports amount to 160 x 10^6 tons. These imports may well increase by 75%, or may double in the next ten years. European and Japanese circles have a strong interest in controlling a substantial share of this oil flow as soon as possible. For instance, Germany would like to control 25% and Japan about 30% of crude imports with domestic companies in the near future (HILL and EALES, 1970). France and Italy already have important national companies and are somewhat independent from the seven major oil companies.

Government to government negotiations are often selected as a medium for gaining control within the oil market. The direct French contacts with governments in Algeria, Iraq and Lybia produced oil contracts which were included in general trade and/or arms agreements. The delivery of the "Mirages" to Lybia would probably play an important part in the French oil policy. The French state-owned oil company, ELF/ERAP, is currently operating in 35 countries. About 75% of their oil production, however, is coming from Algeria. ELF has a share of 51% in another French group: S.N.P.A. (Soc. Nat. de Pétrol Aquitaine).

Another development is illustrated by a practice initiated by E.N.I., the Ente Nazionale Idrocarburi, an Italian state-owned oil company—i.e., the undercutting of the major oil companies by offering modern types of contracts. For instance the E.N.I., together with PHILLIPS, obtained concession rights from PETROMIN, the state oil company of Saudi Arabia, but the contract included the possibilities of participation of PETROMIN after the discovery of substantial amounts of oil. Other contracts in which the European oil companies work only as operator to explore and produce oil but in which the concession and the oil remains the property of the national government have been concluded between the N.I.O.C. and the French E.R.A.P. (ANONYMOUS, 1969 c).

The basic difficulty for an increasing European influence in the world-oil market is the almost total lack of a common energy policy in that area. Even the European Community failed to organize such a policy. In 1964 the member countries of the E.E.C. formulated a Protocol concerning European energy policy but they could manage to include only a few essentials (Protocol 94th session of the European Coal and Steel Community). Recently, new guidelines for European energy policy were adopted based on the mentioned Protocol.

Japanese companies are presently producing in the Neutral Zone in the Middle East, and in Indonesia. Exploration is going on in Qatar, Abu Dhabi, Alaska, Canada and many places in South-East Asia. Exploration is stimulated by the Japanese Government through the Petroleum Development Corporation, formerly conducted by the Japan Petroleum Exploration (JAPEX). The latter company is now engaged in independent operations.

The increasing influence of European and Japanese companies makes the situation in the world-oil market even more complex and more competitive, probably resulting in even greater pressure on oil prices.

The U.S.S.R.

About 70% of the production in the U.S.S.R. is presently coming from the Ural-Emba region, north of the Caspian Sea and west of the Ural. Traditionally the U.S.S.R. is an exporter of crude, although the exported amounts are not large in relation to the world market. Outside the Communist Area, the exports are mainly to western Europe (Italy and Western Germany) and to Japan. Interesting is the price policy of the U.S.S.R. The oil was exported a few years ago to Japan and western Europe for prices ranging from 8 to 9 rouble per ton, while East Germany, Hungary and Czechoslovakia were paying 17 to 21 roubles for a ton (1961-figures from CHAPELLE and KETCHIAN, 1963). In the western countries this is often interpreted as a financial support by the East-bloc countries of exports of the U.S.S.R. toward the West-bloc. This is not true. ADELMAN (1965, p. 94) points out the following: "Much is made, and rightly, of the wide discrepancy between prices to members of the Soviet bloc and prices to buyers outside. That prices to bloc members are always part of a bilateral trade deal may mean that the apparently very high prices are really lower—or really higher. So far, at least, we cannot tell. Such evidence as exists indicates very great discrimination in Soviet export prices. But surely this is the only conduct to be expected of a rational monopolist who meets competitive offers when he must and otherwise exacts a high price. The notion that these high prices "subsidize" and hence make possible the low prices to non-Soviet buyers is unfounded. The sales to non-bloc buyers are well worth making because they are well above costs, and the sales to Soviet-bloc buyers even more worth because they are even further above costs. This is the simplest theory and it fits the facts."

Three years ago the U.S.S.R. decreased the prices for the COMECON-members and increased the prices for the exports to Italy, under political pressure from the east European states (the price for Soviet crude sold in Italy rose from U.S.$ 1.58 per barrel in January 1967 to U.S.$ 1.82 per barrel in December). The developments within the CO-MECON-area from a large factor of uncertainty in the world-oil market. At the end of the next decade, the Soviet-bloc may import 100×10^6 tons per year or more, due to increased consumption (ANONYMOUS, 1970 b). Yet, the developments in the Tyumen-area, east of the Ural, appear promising and may make it possible for the U.S.S.R. to supply again in the next decade most of the COMECON's needs. The increasing contacts between COMECON-countries and Iran, Iraq, Algeria and Syria suggest, however, that the Communist Bloc is being transformed into an importing area (ANONYMOUS, 1968 b).

Suez-crisis 1967

June 5, 1967, saw the beginning of the current Arab—Israeli war, and since that time the Suez Canal has been closed. The war initiated a number of actions directed primarily toward diminishing the influence of the U.S.A. and Great Britain in the Arab countries. The pipelines from Iraq to the Mediterranean were closed for three weeks, and those from Saudi Arabia for three months. Small strikes occurred in the ports of the Persian Gulf; in Lybia the activities of the oil companies were terminated for four weeks. Additional difficulties arose in Nigeria where, from July 5, 1967, the oil exports from the onshore areas in Nigeria ceased due to the civil war.

Despite these difficulties, the oil supply to western Europe was almost unaffected. Iran continued deliveries. Additional exports came from the U.S.A., the Caribbean Area and the U.S.S.R. Additional tanker capacity was organized to cope with the transport problems. The only effect of the crisis was temporarily higher prices for crude oil in western Europe, caused mainly by the increased tanker rates. Stocks in western Europe were in January 1968 even larger than they had been in June 1967!

The Suez-crisis proved that the major oil-companies' policy of spreading risks had been successful, while the industry as a whole showed a remarkable flexibility. The quick resumption of the Arab supplies to western Europe and the U.S.A. was an indication that the revenues from the oil production are for these producing countries too high to be risked in such international difficulties. The help of the U.S.S.R. for western Europe indicated that this country is not as much interested in promoting instability in the world-oil market as was originally assumed by a number of politicians.

Consequently, one of the main arguments for the import restrictions in the U.S.A.—the independence of foreign oil for safety reasons—seemed less convincing.

The U.S.A.

Until now the oil industry has been mother's darling in Washington. Import restrictions kept prices comfortably high for the national producing companies, while the difference in price between Middle East oil and domestic oil was earned by the domestic refiners through quota-allocation schemes (MANES, 1963).

The supply was kept under control by so-called "conservation" measures, where rich oil wells were allowed to produce only for a few days a month, while marginal wells could produce at full capacity. This resulted in a tremendous development of marginal wells to ensure large outputs for individual companies. Already in 1962 Mr. Halbouty, a well-known operator, stated that: "... we have drilled a minimum of 100,000 wells in this state (Texas) that are not needed, at a cost of several billion (10^9) dollars" (ADELMAN, 1965, p. 56). Further, the Federal Government allowed a 27.5% depletion allowance (see last section of this chapter for a closer discussion of the meaning of the depletion allowance)—a type of extra tax write-off of as much as 27.5% of the gross income—to the oil companies.

However, it seems that times are changing for the American oil industry. The depletion allowance has been reduced to 22%. But more important for the world-oil market are new thoughts concerning imports restriction. The cabinet team of President Nixon is in favor of scrapping the import-quota system and replacing it with a tariff system such that the price level would decrease from about U.S.$ 3.30 per barrel to about U.S.$ 3.00 per barrel. These measures would entail significant hardships for the marginal producers in the coastal states (Texas, Louisiana), yet they would probably diminish the negative effects of the "conservation" measures. Such a decrease in price would at the same time enable most of the oil producers of the intra-marginal fields to continue production; it is even possible that total production in the U.S.A. would expand (the total loss of production of the marginal wells would be more than compensated for by the rise in production of the intra-marginal wells, if the conservation measures were completely abolished). Imports, however, would considerably rise.

It is estimated that, if tariffs were abolished completely (an imaginary situation), with a price-fall to about U.S.$ 2.00, imports would equal about 10.5 x 10^6 barrels per day in 1980 or more than 50% of the needs which are estimated to be at that time 20 x 10^6 barrels per day (ANONYMOUS, 1970 c).

Although it is not clear what the future oil policy of the U.S.A. will be, it may be assumed that import policy will become more liberal. Large uncertainties, however, revolve around the developments in Alaska. If Alaska turns out to be a Middle East kind of oil-producing region, with high productivities per well, and if transport problems to the east coast of the U.S.A. can be solved, the oil economics of the U.S.A. may change quite radically.

Future

In the near future, about the next decade, it is likely that the structural world-oil surpluses will continue to exist. If the developments in Siberia and in Alaska prove successful, this will certainly be the case. If, however, these developments are less rapid than expected, the world is likely to be confronted with rising world-oil prices. As usual, the deciding factors in the prognostics for the oil economy as a whole are geology and technology.

Gas markets

The resource base of natural gas in the world is estimated to be 30×10^{15} cubic ft. (cuft.) or 860×10^{12} m^3. About 60% of these reserves are found in the Communist Area, Africa and the Middle East (LAMBERT, 1966). Only a small part of this reserve can be considered proved. In the U.S.A., 690×10^{12} cuft. (20×10^{12} m^3) is considered as proved. About half this amount can be expected to be presently recoverable in the U.S.S.R. Other large proved gas reserves are found in western Europe (120×10^{12} cuft. or 3.5×10^{12} m^3), the Middle East (170×10^{12} cuft. or 5×10^{12} m^3) and North Africa (140×10^{12} cuft. or 4×10^{12} m^3). The world production can be presently estimated at 35×10^{12} cuft. or 1×10^{12} m^3. This production is mainly concentrated in the U.S.A. (about 67%) and the U.S.S.R. (about 19%). Other important producing areas are Canada (4%), E.E.C. (3%) and Roumania (2%) (UNITED NATIONS, 1968). A rapid expansion of the production is foreseen in western Europe. This area will probably be producing 125×10^9 m^3 by 1975.

A key factor in the gas economy is the transport cost. Pipeline-transport costs are high. In the U.S.A. the average costs are U.S.$ 0.14 per 10^3 cuft. per 10^3 miles (TIRATSOO, 1967, p. 353). Until about 1960 this was the only means of transport. Since that date methane tankers came in experimental use for shipments of natural-gas overseas. This new transportation method is still rather expensive.

The reason for the high transport costs of gas in relation to oil is the lower energy content of gas in comparable volumes. One m^3 of oil represents an energy of about thousand times that of one m^3 of natural gas (normal pressure and temperature). Although gas may be compressed to a smaller volume, in practice pipeline pressure does not exceed 10^3 p.s.i. (70 atm.). For transport in methane tankers, 1,000 cuft. of methane is condensed to about 1.6 cuft. under a temperature of $-258°$ F ($-161°$ C). In this case the volume of methane, compared on an energy basis, begins to be comparable to that of crude oil; but due to higher capital costs, the capital ratio for transport of methane: crude is about 4:1.

These transport difficulties have produced a situation in which the major gas-producing areas are presently located near their markets (U.S.A., U.S.S.R., Canada, western Europe). At the same time about 7.5×10^9 cuft. per day, or about 75×10^9 m^3 per year, is flared in the Middle East (UNITED STATES DEPT. OF THE INTERIOR, 1968). This gas is produced along with the oil and cannot be used profitably. Reinjection of the gas is only profitable in solution- or gas-cap-drive reservoirs. It disturbs the production in water-drive reservoirs. Consequently, there exists no unitary world-gas market, but a number of separate ones. Each market has its own price system and its own structure. For instance, the average wellhead price in the U.S.A. is U.S.\$ 0.154 per 10^3 cuft.; in addition to this price come transport and distribution costs resulting in an average consumption price of U.S.\$ 0.516 per 10^3 cuft. The price in The Netherlands at the frontier is U.S.\$ 0.34 per 10^3 cuft. Iran is exporting its gas to the U.S.S.R. for U.S.\$ 0.184 per 10^3 cuft. and in Kuwait the price for local consumption is only U.S.\$ 0.02 per 10^3 cuft.

The supply curve

Clearly, the general characteristics of the supply curves obtained for oil are also applicable to natural gas: in the short term, increasing marginal costs for the individual production units and for the aggregate of a number of gas fields or gas-producing basins, and in the long term probably decreasing costs. The development of new pipeline complexes and methane-tanker constructions can be expected to influence fundamentally the supply curve of natural gas for farther removed consumption areas in the future.

NEUNER (1960) has raised an interesting question: "Is the typical shape of the supply curve not a reason to expect monopolistic attitudes?" The shape of the supply curve is characterized by the several "cost jumps" between, for instance, the marginal costs for the several natural-gas fields. In some cases it is probable that these cost jumps are pronounced. If a company is owner of the low-cost fields in the supply curve and if the demand is relatively inelastic, it can maximize its profits by restricting the output and enforce higher prices. NEUNER (1960) formulates the following "working rule" for maximizing the profits of a company that owns the low-cost fields: "The maximizing price for a partial monopoly is just equal to the costs of an alternative, uncontrolled-supply source". Neuner investigated the markets of the U.S.A. for the possible existence of monopolistic behaviour. He concluded that it could not be proved that such behaviour existed (NEUNER, 1960, p. 279). His main arguments were that the number of companies in control of the low-costs reserves was large enough to avoid monopolistic situations, while the actual development of the price was not as would be expected if maximum monopolistic profits had been the target.

A completely different picture is obtained for the natural-gas discoveries in western Europe. Of the total-proved natural-gas reserves of 3.4×10^{12} m^3, at least 2/3 is indirectly or directly controlled by different combinations of SHELL and ESSO. The N.A.M. (Nederlandse Aardolie Mij) controls (together with the Dutch State) 60% of the Slochteren gas field, the largest onshore gas field in the world, and has an important share in other Dutch gas fields. In Western Germany two companies, BRIGITTA and EL-

WERATH, both SHELL/ESSO combinations, control important natural-gas fields. On the British side of the North Sea the SHELL/ESSO combination operates the western part of the Leman-Bank gas field, the largest offshore gas field in the world. Apart from this SHELL/ESSO has an important share (50%) in the only gas-pipeline company in The Netherlands: the GASUNIE, while the SHELL/ESSO combination has shares of 25% of the important gas-distribution companies RUHRGAS and THYSSENGAS in Western Germany, and a share of 33 1/3% in DISTRIGAS, the gas-distribution company in Belgium. From this evidence it can be easily concluded that the SHELL/ESSO combination maintains a very strong position in the western European gas market. This strong position is based on their rights on the Slochteren gas field. Consequently, the setting of prices for the Slochteren gas had nothing to do with competitive-market behaviour. In fact the price was settled in close co-operation with the Dutch government. Interesting studies have been made on this subject and two of them will be analyzed in some detail.

The gas-price for Slochteren could be established within a rather wide margin, considering the economic framework of energy conditions in western Europe. An anlysis was made by WOLFF (1964) concerning the optimum wellhead price. He proposed the price to be established at such a level that the national-economic value was maximized. He defined the national-economic value as being the sum of the net-present value of the company (SHELL/ESSO) profits, the profits paid directly to the Dutch state, and the consumers' surplus. The consumers' surplus is the positive difference between the amount the consumers are willing to pay and the amount the consumers actually pay. Triangle ABC in Fig. 5 shows this consumers' surplus graphically. Remember that this consumers' surplus is only applicable to the national population; as soon as the gas is exported, the consumers' surplus is exported too, and cannot be added to the national-economic value of the gas reserve. Thus a rather complicated optimalization problem must be considered. The more rapidly the gas is produced, the higher the net-present value (see for the meaning of the concept "net-present value" Chapter IV). But a rapid production can only

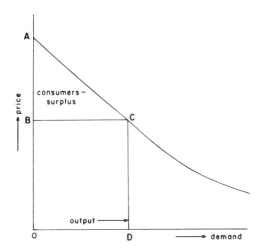

Fig. 5. Illustration of the consumer's surplus

be reached when large exports are included, due to the rather small market in The Netherlands. On the other hand, production must last for a sufficiently long time—assuming that no additional gas reserves can be found—to make the entire change of the energy structure in a country profitable. WOLFF (1964) calculated the optimal price as U.S.$ 0.47 per 1,000 cuft. (or U.S.$ 0.017 per m^3), based on a reserve figure of 1,100 x 10^9 m^3 to be produced over 30 to 40 years.

In this model the reserve was treated as a constant value with a changing price. However, the discussion of the influence of economic factors on the reserve figures has demonstrated that the reserve volume is likely to respond to a changing price. For gas, however, the recovery factor is already rather large (80%) and few additional reserves can be obtained for a single pool when the price climbs higher (the marginal-cost curve is for 80% of the reserve almost horizontal, but for additional percentages almost vertical). If different pools must be included in the model, as is true in The Netherlands, where there is a considerable number of smaller gas fields in addition to Slochteren, the assumption of a "constant" reserve with a changing price cannot be made and the model must include a study of the supply curve.

Another study of the gas-price setting was made by IWEMA, et.al. (1966). They concluded that natural gas could to a great extent replace oil and coal, but that the substitution rate varied widely among different industries, because consumers were forced to make large investments in additional equipment. The price of the natural gas formed an effective medium for regulating the optimal penetration of the natural gas in the total energy-consumption pattern.

The studies of IWEMA, et.al. (1966) and WOLFF (1964) underline the monopolistic character of the production of the Slochteren gas field. The Dutch government made extensive practical use of this inherently monopolistic position. On the other hand, the SHELL/ESSO combination was able to calculate the optimal penetration rate and the optimal price on its own, based on studies that maximized the net-present value of profits. Due to their extensive impact on the European gas market, as previously illustrated, these companies were able to assemble rather complete information for establishing their optimal price (ODELL, 1969).

Quite a different situation exists on the British side of the North Sea gas reserves. A number of different companies own the various different blocks. The direct influence of the state was until recently limited. A key factor in the British pricing system is the influence of the GAS COUNCIL, a body with the sole right to buy the natural gas from companies producing in British parts of the continental shelf. This is a monopsonistic situation for the GAS COUNCIL, which is a state-owned company. It is profitable for the GAS COUNCIL to negotiate for prices that are as low as possible. An extensive study of the British gas market has been made by POLANYI (1967). In his discussion of the problem he assumed an almost completely horizontal supply curve at 1 pence per therm (before devaluation) or about U.S.$ 0.117 per 10^3 cuft., up to a production of 30 x 10^9 therms per year (90 x 10^9 m^3 per year). Above this yearly production the marginal costs would rise slightly to 2 d/therm. On the basis of this supply curve for the North Sea gas, and a study of the demand curve, Polanyi calculates a "monopolist's" optimum price of 2.6 d per therm and a long run competitive-market price of 2 d per therm. The contracts of the GAS COUNCIL presently fix prices at 2.9 d per therm.

It is questionable whether the approximation of Polanyi is applicable, because the production of 90×10^9 m^3 must surely be recovered from a larger number of fields; and since the already discovered fields differ rather widely in proved reserves, distance from shore, depth of reservoir, etc., rather important "cost jumps" may be expected in the supply curve. This irregular shape of the supply curve may alter completely the outcome of Polanyi's analysis.

Following this theoretical discussion of the shape of the supply curve and some of its consequences in a monopolistic-market situation, it is interesting to turn to closer practical analysis of the western European gas market.

The western European gas market

As explained in the previous paragraphs, the Slochteren gas field, and consequently the export policy of the N.A.M. and the Dutch government, play a crucial role in the western European gas market. An interesting debate over the export policy to be applied for the Slochteren gas took place between ODELL (1969), a professor of the Netherlands School of Economics in Rotterdam, and PORTEGIES (1969), member of the marketing department of GASUNIE. The progress of this discussion will be followed and commented upon in the following pages.

The first problem is at what point to fix the depletion rate of the Dutch gas fields. Presently, it is thought that in 1975 the maximum output of about 60×10^9 m^3 will be reached. Since the proven reserves are estimated to be $2,000 \times 10^9$ m^3, the gas will be hypothetically depleted in more than 30 years. According to Odell this is a remarkably slow production rate in comparison with other gas-producing areas. He mentions that the quoted reserve figures normally tend to be conservative. Portegies argued that production decline begins at about a cumulative production of 70% of the recoverable reserves, making it impossible for the production rate to be continued for more than 20 years, which is the minimum time for establishing long-term contracts. Odell's arguments, however, seem most persuasive, since "proved"-reserve figures for the Slochteren field actually mean that the figures have a 5% chance of being too high and a 95% chance of being too low. This "proved-reserve" concept embodies a conservative attitude, and we can assume that the maximum-production rate can thus be continued for a longer time, making the depletion time for the Dutch gas fields indeed excessively long.

The second problem involves the monopolistic character of the present gas market. Odell stresses Slochteren's high wellhead price (being U.S.$ 0.35 per therm)—a price, it must be noted, obtained for a large gas field with excellent reservoir conditions. Prices in the U.S.A. average about U.S.$ 0.15 per therm for numerous small gas fields. Therefore, the rent plus the monopolistic profit for the Slochteren field must be considerable. The monopolistic character of the western European gas market has already been outlined in the previous paragraph. Odell points out further that the monopolistic establishment of the wellhead price must, from the SHELL/ESSO standpoint, surely have included the enormous interests of these companies in other energy sectors, such as the oil industry in Western Germany. A high price for natural gas saves unnecessary competition with regard to these interests. Portegies did not react to these viewpoints.

The main differences in opinion are over the influence of possible competitors for the Dutch gas. The first group of competitors is composed of the already existing and possibly yet-to-find gas fields in the North Sea area. Odell assigns a rather strong weight to the competitive character of the gas from the British areas. Portegies mentions that the discovered British reserves, about 700×10^9 m^3, are relatively small considering the British needs, and does not expect the influence of this gas to reach the continental European gas market. This standpoint seems reasonable in view of the far-reaching activity of the GAS COUNCIL. No definite opinion can be forwarded concerning future reserves, but the discovery of large new gas fields in the North Sea area may lead to greater competition, and consequently a lower export price. This is acknowledged by Portegies.

Odell expects an increase in activity of the U.S.S.R. in the western European gas markets. He cites the large contracts between the U.S.S.R. and Iran, for ultimately 10×10^9 m^3 per year. This gas can be transported to western Europe merely for the cost difference between the investment in a pipeline with a larger diameter instead of a smaller one. Given the low prices calculated on the Iranian border, this gas can be offered in western Europe for U.S.\$ 0.23 per 10^3 cuft. Portegies stresses the political risk and maintains that no European government will seriously consider large Soviet gas imports. The recent important gas contracts between the U.S.S.R. and Italy (6×10^9 m^3/year) and Western Germany (3×10^9 m^3/year) prove that Odell was right and Portegies the contrary. Important gas fields are found in the U.S.S.R. in the Ukrainian and the Ural region, while large discoveries have been made in Siberia. Presently the U.S.S.R. is delivering Italy natural gas for U.S.\$ 0.15 per metric therm, but that is on a market-delivery basis in Milan. The reasoning of ODELL (1969) suggests that even these prices are profitable for the U.S.S.R.

Another point of discussion is the possible influence of the natural gas of Algeria and Lybia. This gas must be delivered by methane-tanker shipments. Restricted shipments are presently made from Arzew—where a liquification plant has been built—to Canvey Island near London, and to France. This gas is derived from the extensive gas field Hassi R'Mel. The production and liquification of this gas is regulated by the Algerian-French contract on a government-to-government basis. New plans are in the offing for methane-tanker shipments from Skikda in Algeria. CONCH INTERNATIONAL METHANE LTD (a combination of Union Stockyard and Transitco: 20%, Continental Oil: 40%, and SHELL: 40%) is also interested in further co-operation with Algeria for more methane shipments. ESSO has important interest in Lybia for shipments from Marsa el Brega to countries north of the Mediterranean. An interesting link may exist between the E.N.I. as an importer of natural gas for the Italian market and its share in the Hewett field in the North Sea. E.N.I. was among the first to accept a low price set by the GAS COUNCIL. This contract greatly influenced further contracts, and strengthened the position of importers!

Odell stresses the importance of technical developments in methane shipment technology and assumes that large quantities of natural gas will soon be transported to all the major ports in Europe. Portegies counters by pointing out the high transport costs of methane shipments. Trends seem to bear Odell out in this case as well. Rapid developments in the long-distance shipments of methane are illustrated by the present contracts for deliveries from Africa to the east coast of the U.S.A., and from Brunei to Japan.

SONATRACH, the Algerian state-owned company and EL PASO, an American company, concluded a contract for liquid-natural-gas shipments in quantities of 10^9 cuft./day from Arzew in Algeria to the U.S. east coast (ANONYMOUS, 1970 d). The natural gas from the West Ampa field of SHELL in Brunei will be transported to Japan from 1972 onwards in quantities of nearly 500×10^6 cuft. per day (ANONYMOUS, 1969 d).

Therefore, the more dynamic view of ODELL (1969) on the future of the European gas market seems preferable.

Questionable, however, is Odell's estimate that in 1975 about 24% of the western European energy market will be converted to natural gas, rather than the generally expected 11%. Latest estimates range between 10 and 14% of the energy consumption (ANONYMOUS, 1970 e).

Yet his general conclusion that the N.A.M.-price must soon be decreased to be competitive with gas from other sources seems reasonable. The price cut of 10% at the Dutch frontier in the beginning of 1969 already points in this direction.

Future developments

If methane-tanker-shipment costs decrease significantly in the next decade—which is highly probable—then the future for natural gas is bright. Natural gas from Algeria, Lybia, and possibly Nigeria will soon be transported in large quantities to Europe; while Iranian and possibly other gas markets in the Middle East will be tapped due to the increasing pipeline technology and transports through the U.S.S.R. The next decade may well witness the rise of a world-gas market. This would be a healthy development, because tremendous amounts of natural gas are now wasted—as oil is produced—in almost all the major oil-exporting countries.

Petroleum law

This section will examine some general aspects of petroleum legislation. Five problems are of importance:
(1)   What is the area of jurisdiction of the petroleum law?
(2)   Who owns the petroleum?
(3)   Which requirements must normally be fulfilled to explore for petroleum?
(4)   Which requirements are necessary for the production of petroleum?
(5)   What are the financial arrangements between the state or landowner and the company?

The area of jurisdiction

When mining legislation is applicable to onshore areas, the geographical extent of the area of jurisdiction can normally be rather easily established, since legislation encompasses a part, or the entire territory, of a state. In most cases the boundaries between the various states are rather well established. Mining concessions, however, are not necessarily restricted to the area of specific national states. For instance, in The

Netherlands a coal-mining concession exists that extends into Western Germany for underground mining—but the area of jurisdiction is considered to be Dutch. Further, between Saudi Arabia and Kuwait is situated a Neutral Zone, where petroleum exploration and production is regulated by both governments.

Offshore the extension of the area of jurisdiction is less firmly regulated. Three different areas can be distinguished: the territorial seas, the continental shelf, and the deep-ocean floor.

The area beneath the territorial seas is normally considered to belong without question to the jurisdiction of the coastal state. The extent of the territorial seas is, however, different for most countries in the world. Japan, Australia, Great Britain, The Netherlands and the U.S.A., for instance, claim a 3 mile zone. Italy has a 6 mile zone, and Mexico a 9 mile zone. Iran, Saudi Arabia, the U.S.S.R. and Venezuela impose a 12 mile zone, while even larger areas are claimed by other nations. The Philippines and Indonesia claim the area among the numerous islands of their archipelagos; and Chili, Peru and Argentina have established a 200 mile zone. Apart from these territorial seas, separate fishery zones are frequently claimed. The large areas claimed by the latter countries are frequently disputed.

More difficulties arise concerning the adjacent area, the continental shelf. In 1958 the Geneva Conference on the Law of the Sea adopted the Convention of the Continental Shelf, which became effective June 10, 1964. This convention regulates the rights of the coastal states on the continental shelf adjacent to their territoral areas. The convention, however, contains a number of inappropriate provisions. The outer edge of the continental shelf is defined as "to a depth of 200 metres or, beyond that limit, to where the depth of the superjacent waters admits the exploitation of the natural resources of the said areas". At the moment that, for instance, manganese nodules (MERO, 1964), become exploitable from the deep-ocean floor, the "juridical" continental shelf would be extended to the deep oceans. If the rules established by the Convention of the Continental Shelf were extended to the ocean floor, roughly half of the areas beneath the high seas can be claimed by the U.S.A., Great Britain, France and Portugal (VAN MEURS, 1967). This would be obviously unacceptable to most of the remaining nations in the world; there therefore seems to be a consensus that a fixed limit should be established for the "juridical" continental shelf (ODA, 1969). This limit can be in terms of the depth of superjacent waters or of the distance from the coast.

Other difficulties arise when several states claim the same offshore area. According to the Convention the boundary between various states must be determined by agreement; and in the absence of agreement, the boundary is the line equidistant from the coasts of the states, unless another boundary is justified by special circumstances. According to ODA (1969), the International Court of Justice has ruled that the equidistance principle is not yet a rule of customary international law. This decision resulted in the necessity for the boundary dispute between Germany, The Netherlands and Denmark concerning the North Sea continental shelf to be reconsidered. An agreement among the three states may soon be expected.

Regulations about the deep-ocean floor do not exist. This may result in considerable difficulties in the near future when new mining techniques facilitate the production of minerals from the deep-ocean floor. For instance, the bottom of certain areas of the Red Sea is covered with rich zinc ores and recently an international consortium, RED SEA ENTERPRISES, laid a claim on these deposits through an announcement in the DAILY EXPRESS, a newspaper edited in Great Britain (ANONYMOUS, 1968a). The U.S.A. has already offered blocks for petroleum exploration and production that cover areas outside the 200 meters depth line—in the Santa Barbara Channel—and earned bonuses from these areas. Territorial claims seem to expand silently.

Ownership of minerals

According to DROUARD and DEVAUX-CHARBONNEL (1966) four different systems of ownership can be distinguished:
(1)   The occupation system.
(2)   The accession system.
(3)   The regalian system.
(4)   The dominal system.

(1)   *The occupation system.* In countries where immigrants settle for the first time, the minerals usually belong to the discoverer or the person or company who claims the mineral area first. Since this system can easily result in disagreements among the various prospectors, the government must enforce rather strict obligations.

(2)   *The accession system.* The accession system existed in the early days of the Roman Empire (FLAWN, 1966). Since access to the minerals is only possible through the surface, the owner of the surface is in this system automatically the owner of the minerals beneath the surface. This system can seldom be applied strictly, because the state must provide rules to assure the safe and reasonable use of the mineral-producing area. This system is applied in the U.S.A. for those areas where the surface is privately owned.

(3)   *The regalian system.* In this case the state reserves the right to regulate mineral exploration and production. The state chooses the persons or companies that are allowed to produce the minerals, and establishes their obligations toward the state. Separate provisions are generally included to compensate the surface owner for possible losses or damage to his property. This system is applied in the French mining legislation.

(4)   *The dominal system.* In the dominal system, the state owns the minerals and is consequently able to require all necessary obligations from mineral companies which propose to produce the minerals. This system is widely adopted in the Middle East and for all offshore areas assessed as under the continental shelf.

In certain countries the latter two systems are clearly to be preferred to the former two, when an orderly development of the petroleum industry is primarily sought. In the U.S.A., most petroleum-exploration companies must spend huge sums on "land departments", on juridical difficulties between petroleum companies and landowners, and on unitization negotiations to "unitize" fields that are owned by numerous companies preventing an orderly and profitable production of the fields.

Exploration provisions

Apart from financial arrangements, several provisions are normally included in an exploration licence. Frequently the right to explore implies automatically the right to produce minerals when minerals are discovered. This is, however, not always the case. The famous Napoleonic Code (1810) distinguished between an exploration and a production licence.

The right to explore a certain area is generally exclusive, which means that only the licensee is allowed to explore for a circumscribed mineral in a given area. For petroleum exploration these exclusive rights are necessary to provide for orderly exploration when the drilling of exploration wells is included in the licence. If only geological surface mapping and geophysical exploration is allowed, non-exclusive licences may be granted.

The concession area for exploration is normally limited geographically and seldom includes the entire area of the state. Further, portions of the area must be relinquished after certain time periods. For instance, 50% of the area must be returned to the state after a period of 5 years. Further, the exploration concession is normally valid for only a limited time period, say 15 years, after which the remaining part of the concession must be returned to the state, if no minerals have been found that can be profitably produced. These regulations are needed to stimulate exploration and to prevent a company from leasing the area for speculative purposes only.

Apart from these measures, exploration activity is normally regulated by establishing working obligations. The company must spend a certain amount of money for exploration per year, or during the duration of the licence.

Finally, provisions are included for regulating bookkeeping methods; all reports and samples must be submitted to the state.

Production provisions

A production licence applies exclusively to a circumscribed mineral. The licence is normally granted for a certain geographical area and the duration of the concession is limited to a few decades.

Work obligations are normally included in the concession provisions, including regulations for safety measures and rules to prevent or compensate damage to third parties, as for instance, pollution resulting from petroleum production.

Similar provisions to those for the exploration licence are included for reports, samples, bookkeeping methods, etc.

Separate provisions are usually made for oil and for gas.

Apart from these general provisions, special stipulations are occasionally included in the contract, such as the obligation to use local or national transportation companies for transporting oil or gas, or the obligation to re-invest profits generated from the production into the economy of the country.

Financial arrangements

Financial arrangements affecting petroleum companies include bonus payments, surface duties, royalties, taxes and participation agreements.

A *bonus* is paid to the landowner or the state to obtain the right to explore or produce in a certain concession area. The bonus may be required at the date the exploration licence is granted to the company. An additional bonus may also be required when a commercial discovery is made, or when production reaches a certain level. For example, in the contract signed between Sharjah and SHELL, an initial bonus is required of U.S.$ 1.5×10^6, and a bonus of U.S.$ 2.5×10^6 when a commercial discovery is made; another U.S.$ 2.5×10^6 must be paid when production reaches 100,000 bbls/day, while, finally, U.S.$ 4.5×10^6 is required when production reaches a level of 200,000 bbls/day (ANONYMOUS, 1969a).

Instead of a bonus fixed in advance the state may grant the concession to whichever company is ready to pay the highest bonus for a concession. In this case the bonus is determined by bidding. If the amounts offered to the government are secret until the highest bidder is established, the system is called *sealed bidding*. If the concession is sold publicly, the bidding is called *auction bidding*. The former system was applied for the companies wishing to explore offshore Louisiana and Texas, and the same system was used during the famous north slope of Alaska sale in 1969 when the companies were ready to pay as much as U.S.$ $900.- \times 10^6$ for the leases.

Surface duties and *rentals* are either fixed yearly payments, or payments assigned according to the amount of the surface occupied by the concession. Surface duties are usually paid to the state, but in the U.S.A. they are sometimes required by the landowner. The amount to be paid may be a fixed annual charge per surface unit, but yearly rising surface duties also occur. For instance, for the Norwegian offshore a surface duty of Kr. 500 per km^2 is required for the first six years, and afterwards the surface duties rise to a maximum of Kr. 5,000 per km^2 per year. In the U.S.A. lease-delay rentals are paid to the landowner. A common price is one dollar per acre per year (CAMPBELL, 1963).

Royalties are paid in relation to the amount of production. These royalties can be expressed as a certain amount of money per volume or weight of petroleum, or as a percentage of the value of the production. For instance, the customary royalty in the Middle East countries before World War II was 4 shillings per ton. Royalties are presently calculated at, for instance, 15% of the posted price for each barrel of oil. Besides being fixed for each produced unit, the royalty may also be calculated according to a *sliding scale*. The percentage is then varied with the output of the concession. For instance, in France no royalty is required as long as the production is less than 50,000 tons per year. Between 50,000 and 100,000 tons 6% must be paid, between 100,000 and 300,000 tons 9%, etc.

Instead of paying a certain amount of money to the state or the landowner, obligations for payments in oil or gas can be imposed. In this case the state or landlord is handling its own petroleum.

Taxes. Most petroleum companies are required to pay the taxes normally applicable in the country to each company and individual. The tax law—especially in most industrialized countries—is rather complicated and includes a large variety of taxes. The most important are excise tax, customs duties, and income tax.

The *excise taxes* include consumption or sales tax applied to a normally large number of goods, and is expressed as a constant percentage of the value of the goods. Within the European Common Market area the value-added-tax system has been adopted.

Customs duties are necessary when goods are imported or exported. There is a tendency toward gradual liberalization of world trade by lowering such customs duties.

Income-tax is applied in most countries. Separate legislation is effectuated for taxing individuals and for taxing corporation profits. The corporate-income-tax regulations differ widely among various countries. It is normally a fixed percentage of the taxable income, although a varying rate may occur such as in Canada where the first Can.$ 35,000.– profit are taxed with a rate of 18%, while the following profit requires 47% tax (HODGSON and BEARD, 1966). The taxable income is calculated by subtracting the operation expenses and the yearly capital costs from the yearly gross proceeds.

For a petroleum company, gross proceeds comprise the proceeds earned through the sale of the petroleum, being oil and/or gas. The price on which the calculation of the gross proceeds must be based is, for instance, in the U.S.A., the sales price in the immediate vicinity of the well (according to MILLER, 1969). In the Middle East, however, this price is the posted price—which may be considerably higher than the realized price—or a special-tax-reference price. Since the petroleum industry is characterized by a high degree of integration, the oil may remain within the company for refining after production. In this case a transfer price must be established that normally equals the sales price or the tax-reference price.

Yearly operating expenses can be deducted from the gross proceeds to arrive at the operating-cash income. From this the yearly capital costs must be deducted. Large differences among the various countries exist as to what may be regarded as capital. Normally yearly capital charges directly attributable to the producing oil or gas fields can be subtracted in all the countries. The amount of "overhead" that may be allocated to various properties is subject to different rules. Outlays for exploratory drilling or bonus payments attributable to a productive lease cannot be deducted in the U.S.A. through depreciation of capital. These costs are regarded as being recovered through depletion allowances.

The *depletion allowance* in the U.S.A. is the permission to deduct a certain percentage (22%) from the gross income to calculate the taxable income. This amount, however, must never exceed 50% of the taxable income before the depletion allowance was subtracted. The percentage depletion, as it is called, is thought to be a compensation for the loss of capital, in which the oil or gas in place is considered as an asset that is consumed during production.

To complicate matters, depletion allowances also exist in other countries, as in Canada. But in this country the depletion allowance is 33 $1/3$% of the net income rather than the gross income. In France a depletion allowance similar to that in the U.S.A. exists, but there 27.5% may be deducted from the gross proceeds. The tax advantages, however, must be re-invested in five years to be recovered definitively by the company. In The Netherlands there is no depletion allowance, but the outlays for exploratory drilling can be deducted for income-tax purposes. From these examples it can be concluded that the income-tax law as a whole must be studied before an evaluation of the attractiveness of the tax law to the petroleum companies can be made.

In the calculation of the taxable income, the royalties and rentals paid to the state or the landowner are normally permitted to be deducted. The bonus can be deducted, must be capitalized, or can be recovered through depletion.

Apart from the mentioned payments, a government or a landowner may require an agreement to participate in the operations. When a state company participates it may be in the form of a *joint venture*. In this case the independent company and the state company form together a new company that explores for and produces the petroleum. An example is, for instance, the joint venture between HISPANOIL and the KNPC (for 60% owned by the Kuwait Government) for a concession in Kuwait.

Participation may also be required after a commercial discovery has been made. Such obligations for *state participation* occur, for instance, in The Netherlands for natural gas, and in most of the new Middle East concessions for oil and gas.

In the U.S.A. the formula of the *carried interest* is occasionally applied in which the "carried" party does not obtain its part in the working interest until the "carrying" party has recovered its investments. This is a form of participation after the pay-out time (see Chapter IV) is reached. A "carried-interest"-type agreement was concluded between Saudi Arabia and Getty-Oil for a concession in the Neutral Zone. Saudi Arabia required a right to participate for 25% after pay-out of investments (CATTAN, 1967, p. 9).

Literature

Petroleum Economics

Adelman, M. A., 1965. The world oil outlook. In: M. CLAWSON (Editor), *Natural Resources and International Development*. Johns Hopkins Univ. Press, Baltimore, Md., pp. 27-125.

Anonymous, 1968a. Technical breakthroughs displayed in offshore drilling. *World Oil*, 167(1): 90-95.

Anonymous, 1968b. Soviet bloc turns to Middle East for oil. *World Petrol.*, 39(10): 46-48, 65.

Anonymous, 1968c. North Sea gas price agreement causes furor. *World Petrol.*, 39(5): 20H-20I.

Anonymous, 1969a. It's expensive offshore. *Petrol. Press Serv.*, 36(1): 27.

Anonymous, 1969b. International trade: O.E.C.D.–O.P.E.C. balance. *Petrol. Press Serv.* 36(9): 349-350.

Anonymous, 1969c. Middle East terms tabulated. *Petrol. Press Serv.*, 36(3): 99-102.

Anonymous, 1969d. Brunei liquified gas for Japan. *World Petrol.*, 40(13): 12 C, D.

Anonymous, 1970a. Capital outlays sharply higher. *Petrol. Press Serv.*, 37(2): 57-59.

Anonymous, 1970b. When oil flows east. *Economist*, 234(6594): 51-52.

Anonymous, 1970c. The American oil dilemma. *Economist*, 234(6601): 62-63.

Anonymous, 1970d. L.N.G.-trade Europe/Africa to U.S.A. *Petrol. Press Serv.*, 37(3): 103-104.

Anonymous, 1970e. Natural gas penetrates Europe. *Petrol. Press Serv.*, 37(1): 7-8.

Arnaud Gerkens, J. C., d', 1962. Aspecten der ontwikkeling van de petroleumtechniek in de laatste 25 jaren, 1. Ontwikkelingen in de exploratiegeofysica gedurende de afgelopen 25 jaar. *Ingenieur*, 74(33): M1-8.

Arps, J. J., 1961. The profitability of exploratory ventures. In: INTERNATIONAL OIL AND GAS EDUCATIONAL CENTRE SOUTH WESTERN LEGAL FOUNDATION (Editor), *Economics of Petroleum Exploration, Development, and Property Evaluation*. Prentice Hall, Englewood Cliffs, N.J., pp. 153-173.

Bradley, P. G., 1967. The economics of crude petroleum production. In: J. Johnston, J. Sandee, R. H. Strotz, J. Tinbergen, and P. J. Verdoorn (Editors), *Contributions to Economic Analysis*. North Holland Publ. Co., Amsterdam, 149 pp.

Cameron, B. A., 1966. The petroleum prospects under the marine areas of the world. In: P. Hepple (Editor), *Petroleum Supply and Demand*. Elsevier, Amsterdam, pp. 23-60.

Chapelle, J. et Ketchian, S., 1963. *U.R.S.S.-Second Producteur de Pétrole du Monde*. Editions Technip, Paris, 314 pp.

Cooper, B. and Gaskell, T. F., 1966. *North Sea Oil—the Great Gamble*. Heinemann, London, 179 pp.

Cronen, A.D., 1969. How a computer can assist well spacing in the North Sea. *Oil Gas Intern.*, 9(5): 50-61.

De Chazeau, M. G. and Kahn, A. E., 1959. *Integration and Competition in the Petroleum Industry*. Yale Univ. Press, New Haven, Conn., 598 pp.

De Wolff, P., 1964. Economische aspecten van het aardgas. *Academiedagen, Centraal Planbureau, 's Gravenhage*, 16: 61-84.

Fanning, L. M., 1954. *Foreign Oil and the Free World*. Mc Graw-Hill, New York, N.Y., 400 pp.

Fisher, F. M., 1964. *Supply and Costs in the U.S. Petroleum Industry—Two Econometric Studies*. Johns Hopkins Univ. Press, Baltimore, Md., 178 pp.

Gibson, R., 1969. The spreading offshore search. *World Petrol.*, 40(2):26-27.

Halbouty, M. T., 1968. Giant oil and gas fields in U.S. *Bull. Am. Assoc. Petrol. Geologists*, 52(7): 1115-1151.

Hartshorn, J. E., 1962. *Oil Companies and Governments*. Faber and Faber, London, 365 pp.

Heller, C. A., 1968. Financial planning for international oil operations. *World Petrol.*, 39(12): 42-43.

Hill, K. E. and Coqueron, F. E., 1966. Oil, a backward glance, a current appraisal, a 10 year look ahead. *Oil Gas J.*, Oct. 17, pp. 72-81.

Hill, R. and Eales, R., 1970. Where the world oil industry is heading. *Intern. Management*, 25(1): 21-26.

Hodges, J. E. and Steele, H. B., 1959. An investigation of the problems of cost determination for the discovery, development, and production of liquid hydrocarbons and natural gas resources. *Rice Institute Pamphlet*, 16(3): 4-153.

Houssiere, C. R. and Jessen, F. W., 1968-1969. Economics of deep well drilling, 1. What it costs to drill below 15,000 ft. *World Oil*, 167(7). 58 62. Idem, 2. Rate of return and payout for wells drilled below 15,000 ft. *World Oil*, 168(1): 65-68, 87. Idem, 4. Outlook for deep drilling. *World Oil*, 168(4): 52 and 55-56.

Hubbard, M., 1967. Oil inside Europe. *World Petrol.*, 38(8): 44-46, 53, 54.

Iwema, R., Klaassen, L. H., De Klerk, A. en Meyer, J. W., 1966. *Energie in Perspectief*. Meyer, Wormerveer, 112 pp.

Kaufman, G. M., 1963. *Statistical Decision and Related Techniques in Oil and Gas Exploration.* Prentice Hall, Englewood Cliffs, N.J., 307 pp.

Lambert, Don E., 1966. Domestic hydrocarbon supplies could supply U.S. for centuries. *World Oil,* 165: 11-14.

Lovejoy, W. F., Homan, P. T. and Galvin, C. O., 1963. Cost analysis in the petroleum industry. *J. Graduate Research Centre, Southern Methodist Univ., Dallas, Texas,* 31 (1/2): 105 pp.

Lovering, T. S., 1943. *Minerals in World Affairs.* Prentice Hall, Englewood Cliffs, N.J., 394 pp.

Manes, R. P., 1963. Import quotas, prices and profits in the oil industry. *Southern Econ. J.,* 30: 1-12.

Martinez, A. R., 1966. *Our Gift, our Oil.* Reidel, Dordrecht, 199 pp.

Mc Lean, J. G., 1968. The transition from domestic to international oil operations. *J. Petrol. Techn.:* 1339-1343.

National Petroleum Council, 1967. *Impact of New Technology of the U.S. Petroleum Industry 1946-1965.* Nat. Petrol. Council, Washington, D. C., 341 pp.

Neuner, E. J., 1960. *The Natural Gas Industry.* Univ. Oklahoma Press, Norman, Okla., 223 pp.

Odell, P. R., 1969. *Natural Gas in Western Europe. A Case Study in the Economic Geography of Energy Resources* (Inaugural Address). Bohn, Haarlem, 33 pp.

Polanyi, G., 1967. *What Price North Sea Gas?* The Institute of Economic Affairs, Hobart, Paper 38 IEA, 55 pp.

Portegies, M. J., 1969. Marktstrategie van het Nederlandse aardgas. *Econ. Statist. Ber.,* 54(2701): 650-655.

Shell Briefing Service, 1967. The capital needs of the oil industry outside North America: 1966-'70. *Shell Briefing Service,* pp. 1-9.

Stekoll, M. H., 1961. Cutting costs of well completions and operations. In: INTERNATIONAL OIL AND GAS EDUCATIONAL CENTRE SOUTH WEST LEGAL FOUNDATION (Editor), *Economics of Petroleum Exploration, Development, and Property Evaluation.* Prentice Hall, Englewood Cliffs, N.J., pp. 129-152.

Stonier, A. W. and Hague, D. C., 1964. *A Textbook of Economic Theory.* Longmans, London, 574 pp.

Symonds, E., 1966. Petroleum trade and payments in transition. *Exploration Econ. Petrol. Ind.,* 5: 221-238.

Tiratsoo, E. N., 1951. *Petroleum Geology.* Methuen, London, 449 pp.

Tiratsoo, E. N., 1967. *Natural Gas—A Study.* Scientific Press, London, 386 pp.

Tugendhat, Chr., 1968. *Oil, the Biggest Business.* Eyre and Spottiswoode, Londen, 318 pp.

United Nations, 1968. World energy supplies 1963-1966. *Statistical Papers, Series J.,* 11 (Sales no. E.68, 7): 1-101.

United States Department of the Interior, 1968. *United States Petroleum through 1980.* U.S. Dept. Interior, Washingrton, D. C., 92 pp.

Weeks, L. G., 1968. The ocean's resources. *Offshore,* 28(7): 39-48, 87, 88.

Wells, M. J., 1968. The Middle East—a year after Suez. *World Petrol.,* 38(8): 30-32.

Petroleum Law

Anonymous, 1968a. Updating ocean bottom mineral activity. *Ocean Industry*, 3(II): 39-42.

Anonymous, 1969a. Middle-East terms tabulated. *Petrol. Press Serv.*, 16(3): 99-102.

Campbell, J. M., 1959. *Oil Property Evaluation*. Prentice Hall, Englewood Cliffs, N. J., 523 pp.

Cattan, H., 1967. *The Evolution of Oil Concessions in the Middle-East and North-Africa*. Publ. for the Parker School of Foreign and Comparative Law. Oceana Publications, Dobbs Ferry, New York, 173 pp.

Drouard, C., and Devaux-Charbonnel, J., 1966. *Legislation Minière Française des Hydrocarbures*. Soc. des Ed. Techniq., Paris, 166 pp.

Flawn, P. T., 1966. *Mineral Resources. Geology—Economy—Engineering—Politics—Law. Rand, McNally and Company, Chicago, New York, San Francisco, 406 pp.*

Hodgson, E. C., and Beard, W. J., 1966. Summary review federal taxation and legislation affecting the Canadian mineral industry. *Dept. of Mines and Techn. Surveys, Ottawa, Mineral Information Bulletin*, 82 MR: 1-25.

Mero, J. L., 1965. *The Mineral Resources of the Sea*. Elsevier, Amsterdam, 312 pp.

Miller, K. G., 1969. *Oil and Gas, Federal Income Taxation*. Commerce Clearing House, Chicago, 509 pp.

Oda, S., 1969. International law on the resources of the sea. *The Hague Academy of International Law, 1969-Session*; pp.

Van Meurs, A. P. H., 1967. Imperialisme op de oceaanbodem? *Socialisme en Democratie*, 24(6): 385-395.

Chapter IV

Decision making in offshore exploration and production

Introduction

Large investments are risked in offshore-petroleum exploration and production. A suitable system of decision-making is, therefore, necessary to ensure that a company arrives at the best decision in view of its capabilities. It is impossible to give a ready solution for solving such decision problems. Each project in petroleum exploration is unique. Each decision to begin exploration, develop a field, or enlarge production-capacity must be made with allowances for many uncertainties.

In the following sections several techniques for decision-making will be discussed. The last section of this chapter will demonstrate how the different techniques can be applied in the various investment decisions.

Decision-making in offshore-petroleum exploration and production is based on the same methods of investment appraisal used in other industries. No special methods are needed. The conventional evaluation procedures, however, must be adapted to the typical aspects of the petroleum industry. These aspects have been discussed in the previous chapters.

The investment-evaluation methods can be arranged according to three criteria. These are:
(1) According to the arithmetic character of the method.
   Again three different approaches are possible:
   (*a*)   the pay-out-time methods;
   (*b*)   the internal-rate-of-return methods;
   (*c*)   the net-present-value methods.
(2) According to the character of the expectation. In this case two alternative possibilities exist:
   (*a*)   methods based on a "best guess" or, in any case a single value for each parameter;
   (*b*)   methods based on a range of values for the several parameters.
(3) According to the assumption about the utility of money.
   It is possible to distinguish between:
   (*a*)   methods based on the constant utility of money;
   (*b*)   methods based on a changing utility of money.
Most investment-appraisal methods are based on some combination of these possibilities. Almost all the methods used in decision-making for offshore exploration and production are based on the constant utility of money. A dollar is worth a dollar irrespective of how the cash flow develops.

In most prospects where risk plays an important role, the methods are based on ranges of values for those parameters which cannot be established with certainty. As soon as the risk is limited and the parameters are known, a single value can be used.

For important prospects the pay-out time, the internal-rate-of-return, and the net-present value are calculated. In such cases all the three figures are useful to the evaluator.

Apart from the cited methods, some different but related investment-appraisal procedures are known in the mineral industry. Some of these will be treated below.

A large part of the chapter will be limited, however, to a closer discussion of the previously mentioned evaluation procedures. First some general assumptions must be dealt with.

It is assumed that the investment(s) and the value of the benefits that will accrue from the investment(s) can be measured as cash flows. This assumption excludes investments that do not generate cash, and many for which it is difficult to measure the benefits in the form of cash. An example is the investment in scientific research, the immediate goal of which is to generate knowledge. In the long run this knowledge can be used to make new products or find new oil reserves but it is often difficult to express this benefit in the form of a cash flow. Most investments, however, which are needed for offshore exploration and production can be evaluated by the study of the cash flow.

The second assumption is that funds for investment are available at an appropriate rate of interest. Problems related to the financing of the different investments are excluded from the discussion. These problems are left to the economists (MERRETT and SYKES, 1962, pp. 348-371). The same is applicable to establishing the rate of return to be used for the net-present-value calculations. BIERMAN and SMITH (1966, p. 144) provide the essentials concerning these questions. Only the use of very high rates of return for handling large exploration risks will be discussed in this chapter.

Finally, before turning to evaluation methods, a brief remark should be made on the type of outlays which are included in evaluations. Normally only those outlays which can be recognized and identified as belonging to the particular investment are included in an evaluation. In the petroleum industry an exception is often made for overhead expenses (GRAYSON, 1960, p. 121). With investments for exploration and drilling projects, overhead expenses are those that make the investment activities possible. Examples of such expenses are those for office space, managers time, research, etc., which cannot be attributed to a particular drilling project for practical or theoretical reasons. These outlays form an important part of a major oil company's budget and must be recovered out of the proceeds from the several investments that are made by the company. Overhead expenses are usually included in the profitability evaluations and are apportioned in one way or another over the outlays attributable to the project.

Arithmetic methods for prospect evaluation

Three different arithmetic methods for evaluating a prospect are possible: with the pay-out time, the internal-rate-of-return, and the net-present value.

The pay-out time can be described as the time lapse necessary to recover with the positive cash flow the initial investment. The internal-rate-of-return is the maximum rate that can be paid with regard to the future cash flow, if the initial investment has been borrowed. The net-present value is the figure that can be obtained by discounting the initial investment and the subsequent cash flow at an appropriate rate of interest. These methods will be discussed more completely in this section after an explanation of the relation between these three methods.

DIEPENHORST (1967a), showed this relation in a simple formula, based on the assumption that the initial investment can be regarded as a negative cash flow. The formula is the following:

$$x = \sum_{t=0}^{t=y} \frac{C_t}{(1+z)^t}.$$

If C_t represents the result of the cash flow in the time period t, the formula can be interpreted in three different ways:
(1)   Assume: $x = 0$
$$z = 0$$

In this case y can be calculated as the pay-out time.
(2)   Assume: $x = 0$
$$y = \text{the lifetime of the project}$$

Now z is the internal-rate-of-return.
(3)   Assume: $y = \text{the lifetime of the project}$
$$z = \text{an appropriate rate of return}$$
The net-present value is given by x.

Although the three methods are related to each other, their discrepancies are striking because they give weight to three different parameters. All the three methods have advantages and disadvantages for the quality of decision-making in petroleum exploration and production.

Pay-out-time methods

A simple investment is an outlay followed by one or more periods of proceeds. The lapse of time required before the outlay is recovered is called the pay-out time. A great number of managers in petroleum exploration are still using as a yardstick that the shorter the pay-out time, the more attractive the investment.

More complicated investments, requiring more outlays, each followed by one or more proceeds, have correspondingly more complicated pay-out periods. For instance, if after the pay-out time is reached additional investments are necessary, it is possible that the cash flow again becomes negative and additional proceeds result in a new pay-out time. In this case the project has two pay-out times, the initial-pay-out time and the project-pay-out time. Such a situation may arise in practice if, for instance, after the installation of the production well and after production has resulted in a positive cash flow, a workover is necessary to make future production possible. A workover is an investment to make oil or gas better flowing to the bottom of the well.

CAMPBELL (1959, p. 447) stated that the calculation of a project-pay-out time is a popular method of evaluating properties for small independent operators, especially when working with short-term-borrowed capital. In this case a pay-out-time calculation is useful to determine whether the operator can repay his borrowed capital in the required time.

Normally small operators do not work on offshore projects. Only major oil companies are able to spend the huge amount of capital necessary for such operations. Pay-out times offshore are usually long, making it impossible to finance offshore operations with short-term capital.

This does not mean that the yardstick of the pay-out time is not one of the methods used in evaluating offshore projects. Pay-out time is, apart form indicating the attractiveness of an investment, also a yardstick for the safety of the investment. In view of the high risks in petroleum exploration (weather conditions, political instability, mis-interpretation of geological facts, etc.) a short pay-out time indicates a relatively safe investment.

The pay-out time method has a few shortcomings. The major one is that the method gives no insight into the development of the cash flow after the pay-out time is reached. It may be that no positive cash flow is generated after this date but it is also possible that large proceeds are earned. The latter situation is clearly preferable to the former, but the pay-out time does not indicate any difference between the projects. A project with a longer pay-out time be may even more attractive than a project with a shorter pay-out time due to the total proceeds recovered. Further, it is impossible to show the absolute amount of outlays and proceeds in the figure of the pay-out time. A very small investment can have the same pay-out time as a larger one.

Additionally in the pay-out-time method, where generally no attention is paid to the time-value of money, it is possible to include a certain interest rate. In this case the z mentioned in the formula of DIEPENHORST (1967a) has a certain value.

For petroleum exploration and production the pay-out-time method has certainly advantages, in spite of its shortcomings, because of the "safety aspect" indirectly inclu-ded in the method. HARDIN and MYGDAL (1968), among others, therefore, underline the popularity of this method in the industry.

Internal-rate-of-return methods

The internal-rate-of-return can be described as the rate of interest which makes the net-present value of the total proceeds equal to the net-present value of the total outlays. The calculation of the internal-rate-of-return is rather laborious. It can be done graphi-cally, by a trial and error method, or with the help of a computer. The result is usually a single figure which can be used as a yardstick to evaluate the project. However, as soon as two or more projects are compared on the basis of their internal-rate-of-return the method reveals an important shortcoming: the implicit and false assumption that money received from a project can be reinvested at the same rate as that of the project. This shortcoming makes the internal-rate-of-return method almost useless in petroleum explo-ration and production. This is because, especially in this industry, the difference in profitability of the projects is determined by nature, and large differences in profitability between the different projects exist.

Another imperfection of this method is that several rates-of-return are possible for one project if the cash flow is somewhat complicated. The z in the formula of DIEPEN-HORST (1967a) may have several values as the sign of the cash flow (positive, negative) changes from one period to another. This type of cash flow occurs regularly in petroleum production, for instance, in acceleration programs. An acceleration program increases the oil-production rate by making additional investments in production wells. The result of this investment is that more oil will be produced at the beginning of the project and less towards the end, since the same total amount of oil is to be produced in both programs. This type of investment gives two internal-rates-of-return.

Although this method is frequently used for investment appraisal in the petroleum industry, it must be stressed that it is not well suited to the evaluation of petroleum exploration and production investments.

Net-present-value methods

The net-present value of a project is the value that can be obtained if all the outlays and proceeds are discounted at a certain rate of interest. This method does not have the shortcomings of the internal-rate-of-return method because the rate of interest to be used is a given fact, based on the economic structure of the company. Further, the net-present-value method provides a single figure for each project. The net-present-value method is a convenient and clear procedure, used frequently with success in petroleum exploration and production.

Procedures derived from the arithmetic methods

Three procedures derived from the arithmetic methods will be discussed in this section: the present-value-profile method, the method of HOSKOLD (1877), and the average-annual-rate-of-return method of ARPS (1958). These methods are for varied reasons important in the mining industry. The first is used frequently by economists; the use of the other two is restricted almost entirely to the mining and oil industry.

The present-value-profile method is in fact identical to the net-present-value method, except that the net-present value is calculated as a function of the interest rate. In Fig. 6, two projects are compared, project A and project B. Project A shows an undiscounted value of U.S.$ 100,000.— and project B has a worth of U.S.$ 500,000.—. The cash flow of investment A, however, is such that the internal-rate-of-return is 20% while project B returns only 10.5%. As long as the interest rate is lower than 9.5% project B is more profitable than project A. As soon as this interest rate is higher the reverse is true. The presentation of the entire curves showing the net-present value as a function of the discount rate is extremely useful if several projects must be compared, because the curves show perfectly the sensitivity of the net-present value to the interest rate. This can be important if the exact discount rate to be used is in doubt. The two projects also show clearly that the internal-rate-of-return method may rate the different projects in another way than the net-present-value method. For a net-present value of 8% project B is to be preferred to project A, while the internal-rate-of-return of project A is higher. WOODDY and CAPSHAW (1960) mention the importance of the present-value-profile method for petroleum exploration and production decisions because the profile gives a better understanding of the rating of the several projects, especially if the cash flow is complicated and multiple rates-of-return occur.

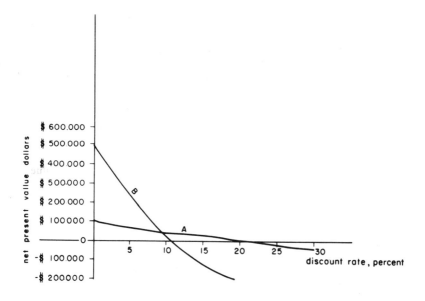

Fig. 6. Illustration of the present-value-profile method

The method of Hoskold was first described in 1877 (HOSKOLD, 1889) and is patterned on the internal-rate-of-return method. The problem of the implicit assumption that the reinvestment rate of the proceeds must be the same as that of the project is corrected by the use of two interest rates: a safe interest and a speculative interest. The method is typically designed to meet one of the characteristics of a mining project: the exhaustibility of the mineral resources. The fact that no exploitable minerals are left if a mine is mined out makes it necessary that the original capital investment must be returned to the investor by the time that the profitable life of the enterprise is ended. To achieve this goal the investor must save each year a part of his earnings and reinvest these savings at a safe interest rate. The savings plus the accumulated interest from the savings constitute a means by which the investor may be able to earn a certain profit. The amount of this profit can be expressed as a percentage of the original investment and this percentage is the speculative interest rate. Appendix I illustrates the derivation of the formula of Hoskold for constant yearly earnings out of a mining property beginning one year after the investment.

The main drawback of the method of Hoskold is that it can be used only for projects with a rather simple cash flow. Furthermore, the method actually shows no more than the net-present-value method, where on the base of a "safe" or any other interest rate a figure for the attractiveness of the project can be obtained.

The average-annual-rate-of-return method of Arps is described by him as follows (ARPS, 1958, p. 339): "The average-rate-of-return computed by this method is essentially the ratio of the present value of the future earnings after amortization to the present value of the undepreciated balance of the investment over the life of the proper-

ty. The method is particularly suitable for investments in oil and gas producing properties, where amortization of the invested capital is on unit-of-production basis." The
method works, as the method of Hoskold, with two interest rates, in this case the bank's
rate and the average-annual-rate-of-return.

The method's main advantage is that it provides a straightforward computation of
the profitability of the project. The main disadvantages are: first, that it loses its simplicity when amortization of the investment is not proportional to the production; and
second, that it cannot be used to calculate the profitability of an incremental investment.

The formula used in this computation is shown in Appendix II, with its derivation.
The use of the unit-of-production depletion is in many cases well suited to the petroleum
industry since most wells have a declining production; consequently the present value of
the depletion is rather high and may be of advantage to the company. The method of
Arps provides in this way a practical procedure for simple petroleum-investment decisions. Its use for other industries, however, is not advisable since the use of the unit-of-production system for depletion may not be applicable.

The three discussed methods derived from the general arithmetic methods are only
a small sample of all the various possibilities that exist in the petroleum industry for
evaluating investments, and many students of the different investment-appraisal methods
would have preferred another selection. The selection of the methods of Hoskold and
Arps, was made for their typical relation to the mining and the petroleum industry.
Before turning to a second grouping of evaluation procedures, a typical relation between
the pay-out time, the lifetime of the project and the internal-rate-of-return must be
discussed.

Relationship between pay-out time, lifetime and internal-rate-of-return

An interesting relationship exists between the pay-out time, the lifetime of the
project and the internal-rate-of-return. If we assume that after the original investment, a
constant stream of equal annual proceeds begins, then pay-out time is easily calculated
from the following formula:

$$\text{pay-out time} = \frac{\text{cost of investment}}{\text{annual proceeds}}$$

If the investment is at year 0, then we know that the present-value of the stream of
equal annual proceeds needs to be equal to or more than the cost of the investment, if we
follow the evaluation by the present-value concept. Thus:
(annual proceeds) x $Art \geqslant$ cost of the investment, or:

$$Art \geqslant \frac{\text{cost of inv.}}{\text{ann. proc.}}$$

In this formula Art represents the value of an annuity for an interest rate r over time
period t.

These two equations lead to the following rule:
$Art \geqslant$ pay-out time (note: Art is dimensionless). This means that the project can be

regarded as profitable if the pay-out time is less than or equal to the annuities for an interest rate r and life t of the project. This rule can only be applied when an investment at year 0 resulted in a series of equal annual proceeds.

If we assume a perpetual life for the project, Art can be written as $1/r$. Consequently, the maximum pay-out time for a profitable project with an infinite life is $1/r$.

This rule also applies when the revenues are not immediately obtained, but begin only after a certain lapse of time. The equation changes in the following way:

$$\frac{1}{(1+r)^n} + \frac{1}{r} = \text{maximum pay-out time}$$

In this formula n is the lapse of time in years between the investment and the start of the equal annual revenues in years.

It is clear that this equation may be applied to petroleum economics because the original investment frequently is followed by equal annual revenues (i.e., contracts for delivery to a pipeline system); the life of the project can be assumed infinite since oil and gas deliveries from the larger oil and gas fields normally last a very long time, for instance twenty years.

The introduction of risk

The methods described in the preceding section did not take risk into account. Petroleum exploration and offshore production is a most risky business, requiring large outlays of money with a high probability of failure. Risk also includes unforeseen adverse weather conditions, accidents, smaller-than-anticipated recovery factors, and often political risks such as wars or nationalization. As soon as the parameters included in the investment appraisal are uncertain, two basically different ways to deal with the situation are possible:

(1) To base the investment evaluation on a "best guess", or in any case to use a single value for each parameter. In this case the procedure can be continued under the assumption that the parameters are known, but with the accompanying knowledge that these parameters are not precise. The previously mentioned three investment-appraisal methods can be applied using the figures obtained from the guess. This procedure is adequate for small and relatively unimportant investments.

(2) To base the investment appraisal on a range of possible values for several parameters. It may be, of course, that several parameters are known. These parameters can be given as a single value or as the known values. The uncertain parameters, however, must be described with a range of possible outcomes. The character of these ranges depends on the type of risk and the type of parameters.

A large part of this section is used to discuss how the different uncertain parameters can be included in the profitability analysis and project evaluation. This will be done for three different types of risk that are important in the petroleum industry: the economic risk, the engineering risk and the geological risk. First, however, it is necessary to consider a question that might arise during a discussion of the handling of risk. That is: why high

interest-rates, combined with the net-present-value method, cannot include risk sufficiently in the analysis.

The use of high interest rates to account for risk

One of the simplest techniques to account for risk is to use high interest rates for
calculating the net-present value. The more uncertain an investment is the higher the
interest rate which is applied. Table VII lists interest rates used by several Canadian
companies in evaluating potential concession areas, as given by HUGO (1965). Rates-
of-return of 25-40% are extremely high. It is questionable whether the application of such
high interest rates gives reliable results, considering the length of time involved between
the initiation of an average offshore-exploration program and the beginning of production. Once production begins, however, the field life may be twenty years or more. The
production twenty years after the start of exploration has no value because of the high
discount rate. It is certainly poor evaluation procedure not to account for production
twenty years or more after the commencement of the project. This is particularly true
where the exploration objective is the giant offshore fields. It is very important in evaluation work to detect and properly take into account a possible giant discovery.

TABLE VII

THE USE OF INTEREST RATES TO ACCOUNT FOR RISK

classification of uncertainty	interest rate required
proven and developed	10%
considered proven but not developed	12 − 16%
probable acreage	25 − 40%

Apart from this shortcoming, the use of high rates of interest makes a special
assumption about the nature of uncertainty. BIERMAN and SMITH (1966) give the
example of an investment to build and equip for producing a new product. In some
instances the major uncertainty may be related to the cost of constructing the plant,
while the demand for the resulting output may be easily predictable, with little accompanying uncertainty. The use of atomic energy to generate electric power is an example
of such a situation. In such a case the discounting of future revenues, themselves fairly
certain, seems a poor way to allow for the uncertainty over how much the fixed plant will
cost.

The objections of BIERMAN and SMITH (1966) to the use of high interest rates
are equally applicable to the investments in petroleum exploration. The lease of acreage
for a new petroleum-production project is similar to the example of the atomic-power
plant.

The uncertainty is mainly whether economically-recoverable oil exists and how large the
reserves are. If oil is economically-recoverable and if markets are available, the uncertainty for selling the oil for a reasonable price is small. Therefore, a high interest rate does
not give proper weight to the element of risk, especially in petroleum exploration.

The introduction of economic risks

Economic risks include uncertainty about future market situations, costs of transport, changes in the future tax structure, etc. Political risks can also be included in this category. In most cases these risks are not insurable. The problems arising in the study of economic risks in petroleum projects are similar to those in other investment analysis. Extensive studies have been made of the forecasting of market supply and demand. It is beyond the scope of this book to discuss the various techniques for handling these problems. An example, however, will be included.

BENELLI (1967) solves the problem of accounting for possible changes in price level by calculating the weighted average of the different expected values. This example is given in Table VIII. The weighted average that can be obtained out of the range of the different values can be incorporated into the normal, previously discussed, appraisal procedures. In the case of the market price such weighted averages can be obtained for the different periods on which outlays and proceeds are calculated.

TABLE VIII

CALCULATION OF THE EXPECTED MARKET PRICE FOR OIL

(1) possible change in market price	(2) price	(3) probability	(2) x(3)	expected price
+ 4%	U.S. $ 3.12	0.50	U.S. $ 1.56	
0%	U.S. $ 3.00	0.25	U.S. $ 0.75	
− 3%	U.S. $ 2.91	0.25	U.S. $ 0.7275 +	
				$ 3.0375

The same procedure can be followed to allow for variations in transport costs. These must be subtracted from the market price to obtain the well-head price.
A higher well-head price can be introduced—due to the "economies of scale"—for a larger volume of gas or oil. The best way of including the economic risk in transport costs is to calculate the probabilities of different well-head prices and to estimate the revenues on a well-head price basis in the same way as proposed for the market price. It is interesting to note that there in an important link between the expected transport costs and the geological expectation of the volume of the reserve, or the economic risk and the geological risk.

Another point is that the positive influence of "economies of scale" in the lowering of the transport costs may be partially or wholly counterbalanced by the possible negative result of pressure from the market that can be expected when the reserve to be produced is large in relation to the market demand (the oversupply hazard). This is one of the economic risks of exploring for a giant gas field on the continental side of the North Sea. DE LAVILLEON (1968) states, for instance, that the gas from such a field cannot be sold for a "Slochteren"—price (about 0.15 U.S.$ cents/metric thermie).

The economic risks in tax structure changes are more difficult to deal with because they are almost unpredictable. Changes in the level of income tax affect the petroleum industry located within the country. These changes are generally due to a change in the overall economic conditions of the country. In politically unstable countries, where tax revenues from the oil industry constitute an important part of the goverment revenues, these changes may be sudden, due to an abrupt change in government policy (MARTI-NEZ, 1966).

Important changes in government policy are also possible in politically stable areas. The discovery of a giant Alaskan oil field, such as Prudhoe Bay for instance, may change the entire oil policy of the U.S.A.

It is almost impossible to foresee developments of this scale and sort twenty or more years in advance. It is this type of economic risk that gives the pay-out-time calculation more value than can be supposed when we look only at the theoretical merits of the method.

The introduction of engineering risks

Engineering risks can be subdivided into those which are insurable, such as damage due to weather conditions, fire, or traffic accidents; and those which are not insurable, such as the imperfection of calculation and measurements. Such imperfections greatly influence production operations, the recovery-factor, other reserve data, etc.

Insurable risks. Offshore exploration and production is still a dangerous business. The construction of offshore drilling rigs and production platforms is a young industry which lacks experience and the necessary technology. In 1954 near-shore work was done in four countries—the U.S.A., Venezuela, Saudi Arabia and the U.S.S.R—under relatively favorable conditions. From that date offshore work has increased rapidly; in 1968 it was going on in 58 countries of the world, including areas with extremely rough weather conditions such as the Cook Inlet in Alaska and the North Sea. Hard lessons had to be learned—as, for instance, following the loss of the "Sea Gem" of the British Coast or the leakage of oil in the Santa Barbara Channel.

It is possible to insure against the loss of equipment, and it is also easy to include insurance costs in the profitability calculations. It is difficult, however, to insure a company against a delay in production or against the claim for a spoiled beach. It is perhaps not realistic to account for risks of this scale within a single project. This does not mean that the risks should not be intensively studied, but that they should not be included in the profitability analysis, because they cannot be assigned to a single project. The company must be able to pay for a catastrophe out of its combined income from all its projects. This is an important reason why only large companies, or large groups of smaller companies, are able to do offshore work. Unnecessary risks can be avoided by eschewing marginal projects—but this is not simple, for many reasons.

Not insurable risks. It is easier to account for uncertainties in measurements and calculations. The best way to demonstrate the use of techniques for handling this type of risks is to give an example described by VAN DER LAAN (1968). The calculation of gas initially in place is made with the following formula:

$$G = V \times O_{av} \times S_{g\,av} \times E_{av}$$

in which:

G = volume of gas in place (1 atm. 15° C);

V = bulk reservoir volume;

O_{av} = average in situ porosity, as a fraction of the bulk volume;

$S_{g\,av}$ = average gas saturation, as a fraction of the pore volume;

E_{av} = average gas expansion factor from reservoir conditions to standard conditions.

These four factors are calculated from a large amount of prior data, but each reservoir is characterized by its own uncertainties.

The simplest calculation for an evaluation is in obtaining a "best guess" for each of the four factors. The "best guess" must be made by qualified people with a sound knowledge of the situation. The multiplication gives a "best guess" for the gas initially in place. This figure can be used for the evaluation of a development or a production program. It is clear that the use of such "best guesses" can be tolerated only for relatively unimportant decisions.

Instead of using "best guesses" it is possible to work with ranges of possible values for one or more parameters. In this case the outcome of the analysis is also a range of possible values. If several parameters show a range of possible values the problem rises as to how these different ranges can be combined in a single distribution.

This problem can be solved mathematically if the ranges for each parameter can be described as probability distributions. VAN DER LAAN (1968) estimated the volume of "gas in place" in the Slochteren-gas fields with a combination of these probability distributions for each parameter. The probability distributions of the basic data could be obtained from a statistical analysis of the data; in other cases they were a quantitative summary of the estimator's opinion. The different distributions for the four parameters in the previously mentioned formula were combined mathematically. The result can be expressed as a cumulative probability distribution; since this distribution is symmetrical, the average expected value for the Slochteren-gas can be read directly from the curve at the point of 50% cumulative probability.

In order to simplify the mathematical work involved in combining probability distributions, an approximation technique may be applied which has also been described by VAN DER LAAN (1968). He classified five values out of the probability distribution:

vo: a very optimistic value, having a change of about 10% of being too low;

o : an optimistic value, having a chance of about 30% of being too low;

m : a middle value or median, dividing the area below the probability curve in equal parts;

p : a pessimistic value, having a chance of about 30% of being too high;

vp: a very pessimestic value, having a chance of about 10% of being too high.

A number of probability distributions for independent variables may be combined as follows. Two sets of five values of two parameters are combined in 25 values. These 25 values are reduced to five again by determining the average of the five highest values, the subsequent five values, etc. The resulting five values are combined with those for a third parameter, etc.

By combining the probability distributions of the four parameters for the reserve calculation, five values for the volume of the gas in place can be obtained, each representing approximately 20% of the total probability range. From these five figures the final cumulative probability curve for the volume of the "gas in place" can be constructed. This curve can be called the *expectation curve*.

When the parameters are characterized by other than simple, normal distributions, the expectation curve is no longer a normal distribution. In this case the "Monte Carlo" technique can be used as described by HESS and QUIGLEY (1963) and HERTZ (1964). This method is especially suitable when not only engineering parameters but also economic parameters must be included in the expectation curve. Such a curve can be, for instance, an expectation curve of the expected net-present value of the project. The method can be easily applied with the help of a computer. HERTZ (1964) describes the procedure as follows:

"To carry out the analysis, a company must follow three steps:

(1) Estimate the range of values for each of the factors and within that range the likelihood for occurrence of each value.

(2) Select at random from the distribution of values for each factor one particular value. Then combine the values for all of the factors and compute the rate-of-return (or present value) from that combination.

(3) Do this over and over again fo define and evaluate the odds of the occurrence of each possible rate-of-return."

An advantage of this method is that very different types of distributions can be included in the procedure. Extremely subjective estimates and exactly defined distributions can both be used. A continuous distribution is as suitable as a simple three point distribution (maximum, average, minimum). Furthermore, the method provides an easy solution for almost all the problems which arise when responsible guesses have to be made. The result of the program is an expectation curve—in the example of HERTZ (1964) an expectation curve of the possible rate-of-return—but a geological reserve may be calculated in this way as well.

Fig. 7 provides a selected example of such an expectation curve. In this case the distribution is given for the expected volume of the "oil in place". From the graph it can be seen that a probability of 0.8 exists that this volume is at least v_1. The median value can be obtained simply by looking at the graph to see which point corresponds to a probability of 0.5 The **average value** can be obtained as the weighted average of the different possible values given in the expectation curve; this value is given as v_a.

If the expectation curve represents the possible outcomes for the net-present value of the project, the average value can be called the *expected-monetary value* of the project. The expected-monetary value is a single value used to rate the attractiveness of the project and can be used as a yardstick. It is better, however, to compare different projects by comparing their different expectation curves, because these curves show the entire range of possible outcomes.

The usefulness of the expectation curves is illustrated in Fig. 8. In this graph two expectation curves are given for two different projects. The distribution of the possible outcomes is in both cases symmetrical, and consequently the average value is the same as the median value in these curves. The median value—in this case the expected-monetary value—is for the two projects exactly the same. The projects, however, are different. Project A shows no outcome with a negative net-present value. Project B, to the contrary,

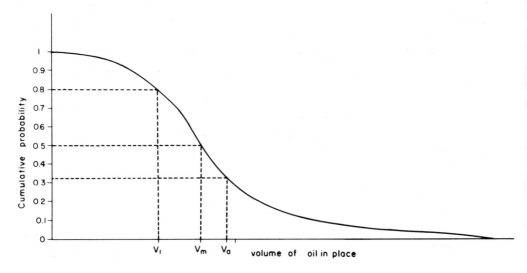

Fig. 7. Expectation curve for volume of oil in place

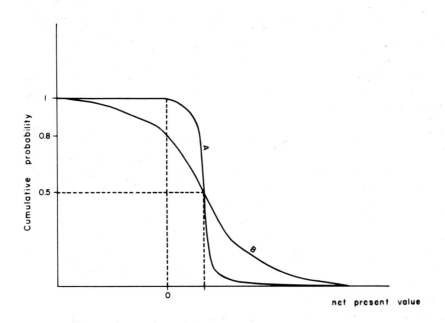

Fig. 8. Two expectation curves showing the advantage of the presentation of expectation curves

has only an 80% chance that the net-present value will be positive. At the same time the possibilities for a large net-present value are greater. It must be management's decision which is to be preferred—yet the two alternatives are in this way better presented.

The methods described in this section can be used for engineering risks and economic risks, or a combination of the two types. One particular difficulty, however, related to the methods of VAN DER LAAN (1968) and HERTZ (1964) must be mentioned. The described procedures must be adapted if two or more parameters are related to each other. For instance, the transport costs of oil are related to the yearly production rate. For each production rate the exact transport costs arc not known but can be described as a range of possible values. In this case the procedure described by HERTZ (1964) must be adapted so that: first, the yearly production is chosen at random; and secondly, out of the range belonging to this yearly production, a certain figure for transport costs can be chosen randomly.

The introduction of geological risks

A major and totally uninsurable type of risk in petroleum exploration is geological risk. The introduction of geological risk into the profitability analysis is not appreciably different from the previously mentioned methods. The geological risk is in fact a type of engineering risk. In view of the very important relation of the geology to the profitability of a prospect, however, this risk is discussed separately.

Geological risk can be divided into two sub-risks:
(1)   What is the chance of finding any oil or gas?
(2)   What is the expected amount of oil and/or gas to be found?

The chance of finding any oil and/or gas is normally expressed in terms of the probability-of-success. It is important to note that this probability is not entirely the same as that used by statisticians. MOOD and GRAYBILL (1963) give the following definition: "If an event can occur in n mutually exclusive and equally likely ways and if n_A of these outcomes have an attribute A, then the probability of A is the fraction n_A/n." A well known example is an urn containing 10 red balls and 90 blue balls. If we are allowed to draw only one ball out of the urn, then the probability of drawing a red ball is 10/100. If the red ball represents success the probability-of-succes in this example is 1/10.

SCHOEMAKER (1963) stresses that what happens in practice is that the evaluator compares the drilling venture in a certain area with such a hypothetical urn with red and blue balls. He believes that his "urn" is filled with so many "red balls" that the probability-of-success in his drilling venture is the same as was the probability of drawing a red ball out of the urn. It must be mentioned, however, that each drilling venture is unique. Therefore, it cannot be simply compared with the hypothetical urn. Further, each wildcat has its own geographical area for which certain information is thought to be relevant. If a wildcat is dry, it is highly probable that each wildcat drilled within a radius of hundred feet from that spot will be dry also. How far away must a wildcat be drilled from another for it to be an "independent trial"? This question presents difficulties, demonstrating the weaknesses of the urn model.

In spite of all these shortcomings, the "probability-of-success" plays an important role in the fixing of the geological risk of a project. Probabilities-of-success are expressed in figures between 0 and 1, or percentages. The absolute figure for this probability-of-succes depends largely on the project.

In the following list the probability-of-success increases generally from top to bottom:

(1)   New field wildcat in an entirely new area.
(2)   New field wildcat in a drilled area.
(3)   New-pool wildcat.
(4)   Deeper or shallower pool test.
(5)   Outpost or extension test.
(6)   Delineation test.
(7)   Drilling for a development well.
(8)   Drilling for a production well.

In this list the differences between the last four categories are very gradual. The probabilities-of-success can be as low as 2% in the upper part of the list or as high as 98% in the lower part of the list.

It is important to define when a drilling venture can be labeled a "success". "Success" is the discovery of oil and/or gas such that the present value of the costs—after the date of discovery—of developing and producing the field is equal to or smaller than the present value of the generated proceeds. The words "after the date of discovery" are important, for the proceeds may be enough to cover development and production costs but not enough to cover the required exploration costs. The outlays for exploration, however, are sunk costs and these outlays can be recovered only out of other ventures if the well is a failure. Thus it is important to regard the investment necessary for development and production as a separate one which can be in itself profitable.

The *determination of the probability-of-success* in the case of a wildcat in an entirely new area is a difficult problem. It is known that the presence of oil and/or gas fields in an area is determined by two factors:

(1)   The presence of a trap.
(2)   The possibility that oil/gas has been generated, has migrated, and has accumulated in this trap.

Prior to drilling, geophysical investigations are normally made to locate possible traps and to provide general information about the new area. It remains for the geologist to determine the probability of finding oil in any of the traps. He normally bases his conclusion on a geological model of the region.

The geologist studies porosity and permeability, the presence of impermeable strata (which prevent the oil or gas from escaping to the surface), and the presence of a favorable environment for the genesis and migration of oil accumulated in the trap.

Each company's exploration department has its own approach to the problem. For the situation in the North Sea, for instance, two different approaches are popular, according to COOPER and GASKELL (1966). One group of geologists follow the "Slochteren" example and looks for large gentle anticlines lying on top of coal measures. Another approach is based on the American salt-dome experience; according to this philosophy it is better to seek locations where more conventional petroleum production from marine sediments is likely. (The first approach was highly successful because large gas fields similar to the Slochteren-field were found off the British east coast).

After a study of all the revelant possibilities, the exploration department makes a prediction of the probability-of-succes.

In a geologically well-known area, where several exploration and production wells already have been drilled, arriving at a probability-of-succes figure is also possible with the help of various statistical techniques. These techniques are based either on the distribution of the known fields, (KAUFMAN, 1963; ARPS and ROBERTS, 1958; McCROSSAN, 1968), or on the urn and ball example (DOWDS, 1968; SCHOEMAKER, 1963). The latter is based on the possibility of making statistical forecasts for the next trial when a sample of red and blue balls is drawn from the urn.

In summary, there are two methods of determining the probability-of-success:
(1)   With the help of geologic models
(2)   Based on the present distribution of oil and gas fields or previous drilling results.

The geological risk involved in the amount of oil or gas to be expected can be determined by using methods such as those described by VAN DER LAAN (1968) and HERTZ (1964). These methods are applicable to the determination of geological as well as engineering risks. Before the drilling of the first exploration well such methods can provide a guess of the volume of oil or gas in place. The "Monte Carlo" technique is especially suited to this purpose. An expectation curve can be obtained for the geological expectation of the volume of the reserves, or any important geological parameter. This curve is constructed, however, from rather subjective estimates of the ranges of the various parameters.

If the expectation curve for the amount of oil and/or gas in a given place is obtained, this curve can be used in another "Monte Carlo" program, together with technical and economic parameters, to construct the expectation curve for the posted net-present value, similar to the curves in Fig. 8 (see p.00) This expectation curve shows the cumulative probability of obtaining at least a certain figure for the net-present value, with the condition that oil or gas actually can be found. Since the probability-of-success is defined as the probability of finding a "successful" production project, a new element can be introduced into the graph. A "successful" production project is one with a net-present value of zero or larger. The pre-discovery investments are excluded from the calculation of the net-present value of the production project. Thus, the expectation curve of the "successful" production project is only a part of the complete expectation curve; it is that part which shows positive net-present values.

In Fig. 9, curve AB is the original expectation curve and curve CB is that part of the expectation curve that provides attractive production projects. It is the latter part of the curve which indicates a "success" and, consequently, the part which is of greatest interest in the evaluation of a drilling project.

Suppose that the drilling venture is a success. In such a case expectation curve CB would represent the cumulative probability of making at least a certain net-present value. In fact, this curve represents the cumulative conditional probability because the stated condition is that the drilling venture is a success.

If, however, there is a chance of only 30% that this venture will be a success, a new

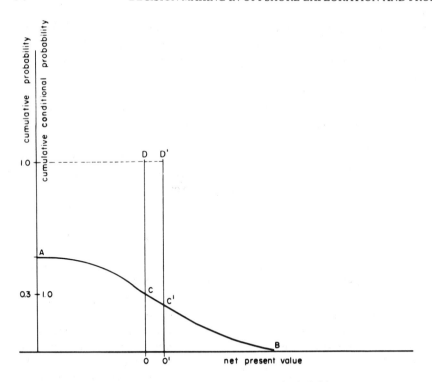

Fig. 9. Expectation curve for a project with a high geological risk.

expectation curve must be constructed: curve DCB. Line segment OC is only 30% of OD. Line DC represents the fact that no successful project can be found, and curve CB represents the expected net-present value. It must be noted that the scale along the vertical axis must be changed during the construction. At first line segment OC showed the cumulative conditional probability; now line segment OD shows the cumulative probability, and C can be found at a cumulative probability of 0.3. It must be remembered that a drilling venture is necessarily a loss as soon as no profitable production project can be found, because the pre-discovery investments, of course, already have been spent.

To determine whether an entire project (including drilling, with accessory dry-hole risk) is profitable, the net-present value of the pre-discovery investments must be included in the calculation. Assume that these pre-discovery investments are represented by line segment OO'. This amount must be subtracted from the expected net-present value of the production project to allow for dry-hole risk. The expected net-present value of a profitable drilling venture can be given as C'B, and profitable drilling projects are found only on the right hand side of O'. The possible losses due to dry holes are given in this final shape of the graph on the left hand side of the line O'D'. Line segment OC represents the probability-of-success and line segment O'C' the probability that an attractive drilling project can be found. The final expected-monetary value can be obtained by subtracting the weighted average of the different values on the left hand side from the right hand side of OD', with the help of the curve DCB.

The expected-monetary value of a drilling venture can be calculated directly when the precision of the expectation curve is not required. CAMPBELL (1962) gives a simple example of how the expected-monetary value can be calculated from a few "guesses" (Table IX).

TABLE IX

CALCULATION OF THE EXPECTED MONETARY VALUE OF A DRILLING VENTURE

(1) possible events	(2) probability	(3) profit	(2) x (3)
dry hole	0.60	U.S. $ − 50,000.−	U.S. $ − 30,000.−
50,000 bbl	0.10*)	U.S. $ − 20,000.−	U.S. $ − 2,000.−
100,000 bbl	0.15	U.S. $ 30,000.−	U.S. $ 4,500.−
500,000 bbl	0.10	U.S. $ 430,000.−	U.S. $ 43,000.−
1,000,000 bbl	0.05	U.S. $ 930,000.−	U.S. $ 46,500.−
	expected-monetary value:		U.S. $ 62,000.−

*) Note that the value of 0.10 cannot be applied to a situation in which exactly 50,000 bbl are found; thus 50,000 bbl represents a range of which 50,000 is the average value.

In some cases it is almost impossible to arrive at a satisfactory figure for the probability-of-success. When this happens, the expected-monetary value can be calculated for a number of different "probabilities-of-success", and the expected-monetary value can be expressed as a function of the probability-of-success.

The exploration department may be able to tell later whether the probability-of-success is larger or smaller than one which makes the expected-monetary value zero.

The two questions:

(1) What is the chance of finding any oil or gas?

(2) What is the amount of oil and/or gas that can be expected?

can also be solved when no exact figures at all can be given.

It is possible to rank different investment projects according to several parameters and to group them in several classes. GRAYSON (1960) gives some examples of these group-rankings. For a well-developed oil country the following classes can be devised for the degree of risk for the different ventures:

(1) Class I ventures are those where there have been nearby (1-2 miles) "shows" of oil and where reliable geological data are available.

(2) Class II are those where there are adequate geological data, but in which certain questionable hypotheses must be made.

(3) Class III are the wild ventures.

It is of course possible to make several types of group-rankings for different problems.

The utility concept

The third method of distinguishing among investment appraisal procedures is according to their use of the utility concept.

The preceding paragraphs have discussed the techniques of arriving at a suitable figure for decision-making, when comparing several projects. For most investments in petroleum exploration and production, risk is a key factor.

Although risk can be expressed in exact figures, the reader must remember that these figures have different meanings for different people. A "gambler" is willing to accept a larger risk than someone who prefers safety in his investments. In many instances it is difficult to say why the one is ready to accept a certain risk while the other will not. In other cases, however, logical explanations present themselves for the differences in attitude. One such factor might be the relative weight of a possible loss compared with the total monetary worth of a company. This difference in value appraisal represents a difference in utility for a particular person or company.

The utility of different values can be given in utility figures. These are figures that express the utility of a certain profit or loss for a person.

The different methods of investment appraisal can be divided into two groups:
(1)   Those that use simply the value expressed in money.
(2)   Those that use utility figures.

It should be evident that all the previously described methods belong to the first group, because they all simply used money as the basis of the calculation.

The following short section, on the other hand, will illustrate the use of utility figures.

After utility figures have been obtained, projects can be evaluated with the same techniques previously presented, with the utility figures substituted for money. Table X shows the calculation of the expected utility of the same project for two different persons, if the utility figures are known. The technique for obtaining the utility figures is

TABLE X

ARBITRARY EXAMPLE FOR THE CALCULATION OF THE EXPECTED UTILITY OF A PROJECT FOR TWO DIFFERENT PERSONS

Table A   utility figures for two different persons:

expected monetary value	person A	person B
− U.S. $ 50,000.−	− 100	− 20
U.S. $ 0.−	0	0
+ U.S. $ 200,000.−	+ 80	+ 60

Table B   calculation of the expected utility

(1) utility		(2) probability	A (1) x (2)	B (1) x (2)
A	B			
− 100	− 20	0.60	− 60	− 12
+ 80	+ 60	0.40	+ 32	+ 24
			− 28	+ 12

Note: the project seems unprofitable to person A and profitable to person B.

extensively described by GRAYSON (1960). In practice, these figures are seldom employed. Most authorities hold that the figures provide little additional information about a project because of their subjective and unstable character. It is interesting to hypothesize, however, whether the method could possibly be reversed. The utility figures could first be established for the company, based as far as possible on rational criteria. Secondly, these figures could be used in the final calculations. In several circumstances such a method might save a great deal of managers' time.

Attitudes towards risk

As soon as considerable geological risk must be taken, two questions of importance arise for the petroleum company:
(1)   Is the company able to absorb a possible loss?
(2)   What are the possibilities for diminishing the company's exposure to the risk?
These two questions will be handled in more detail in this section.

The ability to absorb a possible loss

It is clear that the larger the capacity for investment in relation to the amount of the pre-discovery investment, the better able the company is to absorb a possible loss.

A frequently applied calculation is the chance of *gambler's ruin*. This is the chance that repeated, similar pre-discovery investments do not result in final success. For instance, a company might able to invest in five similar ventures for the discovery of oil. In each of the five cases the probability-of-success is 10%. Thus, there is a chance of 90% that the hole will be abandoned. The chance to drill five wells that must be abandoned is $(0,9)^5 = 0,59$ or 59%. If the company has spent the money in five dry holes there is no possibility for further investments and the chance is 59% that the company is "ruined".

The probability of finding other combinations of events is given in Table XI. It follows from this table that the chance of a gambler's ruin can be expressed as: $(q)^n$. In this formula q represents a figure between 0 and 1 illustrating the probability of a dry hole, and represents the number of similar pre-discovery investments that can be made by a company. The more capital available for exploration, the more similar trials that can be

TABLE XI

PROBABILITY OF THE OCCURENCE OF DIFFERENT EVENTS IN A DRILLING PROGRAM, BASED ON A PROBABILITY-OF-SUCCES OF 10%

events	probability
5 dry holes	$(0.9)^5$ $= 0.590$
4 dry holes — 1 success	$5 \times (0.9)^4 \times (0.1)$ $= 0.328$
3 dry holes — 2 successes	$10 \times (0.9)^3 \times (0.1)^2 = 0.073$
2 dry holes – 3 successes	$10 \times (0.9)^2 \times (0.1)^3 = 0.008$
1 dry hole — 4 successes	$5 \times (0.9) \times (0.1)^4 = 0.000$
5 successes	$(0.1)^5$

done and the lower the chance for a gambler's ruin. The calculation of the chance for gambler's ruin shows that those companies who can afford a large number of independent trials have a smaller chance of being ruined.

In practice, a number of additional factors must be considered.

First, the probability-of-success (p), or the probability of a dry hole $(1-p=q)$, is in most cases different for each drilling project. Secondly, all drilling projects require different pre-discovery investments because they are never completely similar. Finally, the trials are never independent, because after each drilling project results are evaluated; thus the guess of the probability-of-success is frequently influenced by the results of the previous drilling projects in the same area, because this probability figure is based on an extensive study of the available geological and geophysical facts. The "drilling success tree" that occurs in reality is, therefore, somewhat different from the "tree" for the calculation of the gambler's ruin. An example is given in the two "trees" on Fig.10.

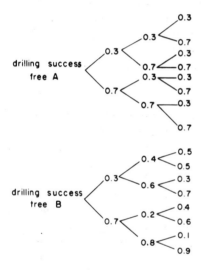

Fig. 10. Two drilling-success trees

Decision tree A is based on the assumption that the probability-of-success is not influenced by the other trials, as is assumed in the "gambler's ruin" calculation. Decision tree B is based on the assumption that a geologist is normally rating the probability-of-success higher when there are already positive results in the prospect area and lower when there have been negative results. The possibilities for several different combinations of events are given in Table XII.

Different trials are dependent when the area of study is such that geological or geophysical factors can influence the probability-of-success in the group of proposed drilling projects in the area. When these factors work in concert it must be assumed that one negative result increases the chance of more negative results; likewise one positive result increases the chance of more positive results. There are numerous factors in such

TABLE XII

PROBABILITIES OF FOUR DIFFERENT EVENTS BASED ON THE DRILLING-SUCCESS TREES OF FIG. 10

events	tree A	tree B
3 dry holes — 0 successes	0.343	0.504
2 dry holes — 1 success	0.441	0.266
1 dry hole — 2 successes	0.189	0.170
0 dry holes — 3 successes	0.027	0.060

areas which influence the possible presence of oil and/or gas in all the traps in the area. These include: the presence of impermeable strata, the presence of good porosity and permeability, and the presence of conditions in which hydrocarbons can be generated, can migrate and be entrapped. Such factors do not necessarily influence the probability-of-success of all the drilling projects in the same way and to the same extent. In fact, there are possibilities that the absence or adverse nature of seferal factors in one part of the area increases the attractiveness of other parts of the prospect area. Poor porosity, for instance, in one of the drilled holes may suggest good porosity in a geological model for the next project, in an adjacent area.

Taking the previously mentioned factors into account, the practical geologist will recognize that exact percentages do not suit his purposes and that these calculations are only significant in obtaining an order of magnitude.

The possibilities of reducing the exposure to risk

The first possibility for reducing the exposure to risk is, of course, a better study of the geological model and the geophysical and geological facts. This study may enlarge the likelihood of choosing the best drilling site.

A substantial reduction of risk can be obtained by an appropriate combination of various investment projects: the portfolio problem. The proceeds generating from different investments can be influenced by the same event. The discovery of a large oil field in a particular area will undoubtedly swell the proceeds from investments in connected transport facilities which must move the newly discovered oil. The same discovery might diminish the proceeds from other less prolific oil fields in the same area through the diminishing demand for their oil.

A company is able to invest in several ventures which are influenced differently by the same events. Such combinations can considerably reduce the risks inherent in exploring for and producing oil. For instance, assume that there are two regions in a certain area. The first is an oil-producing region and the second is untested. A company is able to spend U.S.$ 20×10^6. Three investment projects are possible, each of U.S.$ 10×10^6. Two projects are exploration projects in the untested region; (projects A1 and A2), the third is a secondary-recovery project in the oil production region (project B). The three projects are influenced by the same possible event: the discovery of oil in the unexplored region. Since the company is able to spend U.S.\$ 20×10^6, only two projects can be selected. What is the best combination of projects?

The net-present value of the investments for the two events:

(1)　e_1 the absence of oil;

(2)　e_2 the occurence of oil;

is presented in Appendix III.

The absence of oil in the unexplored basin may affect project B in a favorable way. The demand for oil in the producing region may rise and the proceeds from the secondary-recovery project may be higher. If oil is discovered, the secondary-recovery project is still profitable but only shows a present value of U.S.\$ 1×10^6. Geologists guess the probability-of-success for the exploration project to be 10%. The expected-monetary value for the three projects is given in Appendix III. The expected-monetary value of the three projects is identical, but the exposure to risk in which the company becomes involved is widely varied. Since A1 and A2 are totally comparable, there are only two different choices of projects: A1 + A2 or A1 + B (= A2 + B).

The expected-monetary value of the two different choices of projects is the same: U.S.\$ 20×10^6. The combination A1 + A2 shows the possibility of a loss of U.S.\$ 18×10^6, but also the possibility of a profit of U.S.\$ 38×10^6. The combination of A1 (or A2) + B however, is financially riskless in (this simplified example) because the loss of the drilling project is offset by the extra proceeds from the secondary-recovery project. The two combinations have, as can be expected, the same monetary value.

It depends upon the utilities of the different profits and losses which of the two combinations should be selected. The profitability of project B reacts in an opposite way on the events e_1 and e_2 from the project A1 and A2.

If a company concludes to invest in B and A1 or A2, B can be called a *defensive investment* with reference to the exploration ventures A1 and A2. The strength of major international oil companies is due largely to the extensive possibilities for these defensive investments in exploration, production, transportation, refining, and marketing of petroleum and related products.

For small companies there are also possibilities for reducing the exposure to risk. Several companies can combine their investments, with each taking a share in the total outlay. On an extensive scale this is the principle on which the stock market is based. Companies may also find it profitable to combine investments in exploration projects. This is especially beneficial offshore, where the problem of large risks is severe. Even major oil companies take part in this *pooling of risk.*

The risk-sharing can be extended over several projects in this way for the same investment possibilities. An interesting development is taking place in the U.S.A. According to ROGERS (1969) independent oil companies, as King Resources Group, sell oil and gas "funds" publicly to finance operations. With these "funds" companies generated in 1968 more than U.S.\$ 500×10^6.

The principles of the defensive investments and the pooling of risk are widely known also outside the petroleum industry.

Apart from these measures for reducing the exposure to risk, another aspect of the exploratory ventures, typical to the petroleum industry, should be mentioned. A single company seldom leases the entire offshore of a country for exploration. In most cases the government leases blocks of the offshore to different companies. This gives the companies the opportunity to follow each other's exploration activity. The results in neighboring offshore blocs may give additional information about the geology, and the oil and gas possibilities of the blocks possessed by the company. A chess-board pattern of concessions in an offshore area diminishes the geological risk without formally organizing the pooling of risk by the various companies. Such a patch-work division of an offshore area is common in most of the important offshore areas, as in the North Sea and the Gulf of Mexico.

The decision-making procedure

The previous sections in this chapter have presented the decision-making procedures common in the petroleum industry. Many different methods and approaches were described. This section will apply these methods to the various problems that companies may encounter during actual exploration and production phases. Decision-making is an almost continuous process for a petroleum company, but it is possible to distinguish five distinct phases. These are:

(1)   The decision to start exploration.
(2)   The decision to lease acreage.
(3)   The decision to drill a test well.
(4)   The decision to develop the field and to start production.
(5)   The decision to enlarge production.

The five phases are usually present in petroleum projects and are placed in chronological order. The geological risk is the greatest during the first phase and becomes less during each successive phase. This influences strongly the decision-making procedures that can be applied during each phase. The decision to begin exploration is normally based rather firmly on the geological expectation, although other factors may be of importance. The decision-making procedures used to decide whether production must be begun or enlarged, do not differ essentially from procedures used outside the petroleum industry.

This section will describe the use of various decision-making procedures, particularly during the first three mentioned phases. The passage will also highlight some pertinent literature in the field, and specialized evaluation procedures not described in the previous sections.

The decision to begin exploration

Major oil companies are normally working in several areas at a given time. A new exploration program can be plotted in a "new area" or in an "old area". Generally the major portion of the funds available for exploration is allotted to areas where the company already has experience and is able to make qualified decisions. The remainder is spent in "new areas". The new areas can be divided into those which have been already explored by other companies, but are "new" to this company; and those which have not yet been explored by any company.

First projects will be examined where no specific prior knowledge, garnered from former exploration or production is available. This is still the case for many of the offshore areas of the world. The decision to begin exploration in a new offshore area is based on geological data, and equally important engineering and economic information. Engineering factors include water depth, weather conditions, and availability of repair and supply companies. Economic factors include the availability of markets and the economy of transport. Not to be forgotten are existing oil legislation and the general political situation.

An untested area. If the area is untested two different ways to analyse the decision are open:
(1)   With a geological analysis.
(2)   With a statistical analysis.
        Two examples can be given of a *geological analysis.*
        The first is the method of WEEKS (1965), which is described in Chapter II of this book. This method gives figures only for overall possible reserves of the basin. The positive aspect of the method of Weeks is that a general decision can be made whether an offshore area is promising enough to justify expenses for obtaining more detailed information.
        The second example is given by KÜNDIG (1962) and is based on a rating of the different areas according to several parameters. Kündig specifies a number of such parameters which can be used by geologists to classify different areas according to the oil and gas possibilities. The rating principle is useful if the exploration activity must be limited to a few areas out of many possibilities. The rating principle cannot be used, however, to determine whether exploration may commence or not, because the areas are rated relative to each other and the best area may still be unattractive to explore. The decision to begin an exploration must be made on a more explicit average expectation, as is given by Weeks, of the absolute amount of oil and/or gas anticipated.
        HANSSMANN (1968) describes a statistical approach based on work of ALLAIS (1957) designed to meet the problem of deciding whether a large-scale exploration project in an unknown area should be begun. He studied the problem for the Sahara Desert with regard to ores. His task was to determine whether a large-scale explorarion program in this area would result in the discovery of sufficient profitable ore reserves to justify the expenditures for the exploration. It was found that the value of the annual production of mineral deposits for several statistically well-known areas in the world showed a log-normal distribution. The median values, and the standard deviations, were comparable, so application of the results to the Sahara project was possible. A detailed description of this method is available in the referred to work. Its value lies in the clear picture it presents of the possibilities which may emerge. Such a study falls short, however, in giving no clue to the optimal exploration approach to be followed to pinpoint the few very rich deposits which alone can be profitably mined in remote areas.

        A tested area. The decision to begin exploration in a partially explored area can rest on geological information and on statistical studies as proposed by KAUFMAN (1963), or ARPS and ROBERTS (1958). When several producing fields are known, the possible distribution of the ultimate known reserves can be obtained from statistics. The distribution can be log-normal. When it is assumed that the distribution of all the fields presently known and unknown is also log-normal, this original log-normal distribution can be guessed (ARPS and ROBERTS, p. 2563). The difference between the curve of the known fields and the guessed, original, log-normal distribution of known and unknown fields expresses the chance of finding additional oil in fields of a certain size in a certain area. The difficulty of this method lies in the estimation of the original log-normal distribution. Inherent in this method is the fact that the chance for the discovery of large additional oil fields is relatively smaller (in relation to the present distribution) than the chance of finding a smaller field.

DAVIS (1968) demonstrates this fact with statistics. Five of the major oil basins in the U.S.A. (Los Angeles Basin, San Joaquin Basin, Big Horn Basin, Permian and Texas Gulf Coast) contain 123 giant fields. Of these fields, 32% were discovered within five years of the start of the exploration in the basin, 43% ten years after the start, and 82% twenty years after the start of the exploration. This stresses the importance of the first years of exploration in a new basin. This conclusion is especially important for the new phase in the search for petroleum offshore exploration.

Apart from the mentioned methods the Monte Carlo technique can be used to evaluate the exploration venture.

The decision to lease acreage

The decision to lease acreage is largely determined by existing offshore-oil legislation. Normally it is necessary to obtain a permit or concession before the coastal state permits the drilling of a test well. To obtain this acreage, a bonus usually must be paid. This bonus can be a constant amount per square mile, or it may be necessary to apply the system of sealed bidding.

The decision to lease acreage is made after sufficient results from the geophysical exploration are available. For each block, an expectation curve of the hypothesized amount of oil and/or gas is drawn and the probability of success is established, as explained in previous sections in this chapter. After the results of the evaluation are known, the attitude towards risk is established. Project-evaluators take into account the total strategy of the company. The necessity of pooling the risk with other companies is studied.

When the bonus is a constant amount of money known in advance it can be included in the calculation of the expected-monetary value. When the bidding system is required, the acceptable upper limit of the money which can be allocated to bidding must be calculated.

Normally top management decides the maximum sum which can be spent for bidding in a certain area. The higher the bid for a certain block the greater the chance for making a winning bid. But the higher the bids for several blocks, the less the total amount of bids that can be made for the specified amount of money. To choose an optimum bidding strategy is a complicated problem.

HANSSMANN (1968) and ARPS (1965) propose contrasting solutions for this problem. ARPS (1965) supposes that the different bids on the same tract show a log-normal distribution. He introduces a figure for the bidding strategy (S), which indicates the bid value in terms of the expected-mean-bid value. Past experience shows the relation between S and the chance of being a high bidder. Employing this function he devises an optimum strategy (S) for the bidding on a special block.

HANSSMANN (1968), on the contrary, concludes from previous bids on similar areas that the winning bids of the *different* tracts show a log-normal distribution.
He separates the new blocks into four groups on a scale of attractiveness, each having a relative value one to the other. He then calculates the possible log-normal distributions for the winning bids for these four groups. He approaches the optimum bidding strategy by calculating the marginal gain as a function of the height of the bid for each class, and by combining different sets of blocks (with constant relative values to each other). The method of HANSSMAN (1968) depicts more clearly the relation between the different bids on different blocks by showing the optimization process for an optimal bidding strategy for several blocks together.

With a specified amount of money it is possible to bid with two different strategies:
(1)   To bid for a few very good blocks
(2)   To bid for many blocks of medium quality

The merits of a good block are usually recognized by several companies, and bids are correspondingly high. Consequently, much money is dispensed to obtain few acres. When medium-quality blocks are required, it is sometimes possible to obtain many units for a similar amount of money. The two strategies became clear after bidding for the May 1968 offshore-lease sale in Texas. Much of the bidding was based on about U.S.$ 30×10^6 worth of seismic work conducted by twenty firms. Among these were Texaco, spending U.S.$ $183,3 \times 10^6$ for twenty tracts and the Alamos group spending "only" U.S.$ 18×10^6 for 39 tracts and 15 quarter sections. It is still too early to conclude which of the strategies was the best.

The decision to drill a test well

The decision to drill a test well includes a considerable amount of geological risk. The decision can be based on the expectation curve obtained from the estimate of the probability-of-success, and the expectation curve of the net-present value expected. These expectation curves can be obtained by using the Monte Carlo technique.

The decision to drill a test well can be delayed when it is beneficial to await results from nearby blocks. If the outlays for drilling are expected to be extremely large, as in the North Sea where rough weather conditions prevail, considerable (relatively cheap) seismic work must be done before the test well is drilled. The availability of sufficient seismic data of good quality reduces the risk of drilling a dry hole and consequently enlarges the expected-monetary value of the drilling projects. The true economic meaning of the seismic work is to increase considerably the probability-of-succes and to enlarge the chance of finding a gigantic oil and/or gas field. Offshore work is economical only when fairly rich deposits can be found and these deposits are located generally in large structures. This intensive seismic activity leads, according to WEEKS (1968), to higher wildcat-success ratios (a posteriori) offshore in comparison with onshore ratios.

When a company has acquired several blocks, an ideal spending pattern must be calculated. This can be done with the help of linear programming. TIMM (1969) discusses this method in regard to exploration plans.

The decision to develop the field, begin or enlarge production

The decisions to develop the field and to begin or enlarge production are made with the help of the procedures described previously in this chapter. These procedures are not essentially different from procedures used outside the petroleum industry and it is of little interest to examine details of their exact application.

Literature

Allais M., 1957. Method of appraising economic prospects of mining exploration over large territories. *Management Sci.* 3: 285-347.

Arps, J. J., 1958. Profitability of capital expenditures for development drilling and producing property appraisal. *Trans. A.I.M.E.*, 213: 337-344.

Arps, J. J., 1965. A strategy for sealed bidding. *J. Petrol. Technol.*, 17: 1033-1039.

Arps, J. J. and Roberts, T. G., 1958. Economics of drilling for Cretaceous oil on east flank of Denver Julesburg Basin. *Bull. Assoc. Petrol. Geologists*, 42(11):2549-2566.

Benelli, G. C., 1967. Forecasting profitability of oil-exploration projects. *Bull. Assoc. Petrol. Geologists*, 51(11): 2228-2245.

Bierman, H. and Smidt, S., 1966. *The Capital Budgeting Decision. Economic Analysis and Financing of Investment Projects*. Mc Millan, New York, N.Y., 420 pp.

Campbell, J. M., 1959. *Oil Property Evaluation.* Prentice Hall, Englewood Cliffs, N. J., 523 pp.

Campbell, J. M., 1962. Optimization of capital expenditures in petroleum investments. *J. Petrol. Technol.*, 14: 708-714.

Cooper, B. and Gaskell, T. F., 1966. *North Sea Oil—the Great Gamble.* Heinemann, London, 179 pp.

Davis, L. F. 1968. Economic judgement and planning in North American petroleum exploration. *J. Petrol. Technol.*, 20: 467-475.

De Lavilleon, P., 1968. Adriatic search is underway. *World Petrol.*, 39(3): 48-50.

Diepenhorst, A. I., 1967a. Investeringsselectie in theorie en praktijk. *Maandbl. Accountancy Bedrijfshuishoudkunde*, 415: 165-174.

Diepenhorst, A. I., 1967b. *De Portefeuillekwestie* (Rede uitgesproken op de vierentwintigste dies natalis der Nederlandse Economische Hogeschool op 8 november 1967 door de Rector Magnificus). Nederlandse Economische Hogeschool, Rotterdam, 20 pp.

Dowds, J. P., 1968. Mathematical probability approach proves successful success ratio during 1964-'68 averages better than 50% for 118 wells, including wildcats and extensions. *World Oil*, 167(7): 82-85.

Grayson, C. J. jr., 1960. *Decisions under Uncertainty. Drilling Decisions by Oil and Gas Operators.* Harvard University, Graduate School of Business Administration, Boston, Mass., 402 pp.

Hanssmann, F., 1968. *Operations Research Techniques for Capital Investment.* Wiley, New York, N.Y., 269 pp.

Hardin, G.C. jr. and Mygdal, K., 1968. Geologic success and economic failure, *Bull. Am. Assoc. Petrol. Geologists,* 52: 2079-2091.

Hertz, D. B., 1964. Risk analysis in capital investment. *Harvard Business Rev.,* 42(1): 95-106.

Hess, S. W. and Quigley, H. A., 1963. Analysis of risk in investments using Monte Carlo techniques. In: AMERICAN INSTITUTE OF CHEMICAL ENGINEERING (Editor), *Statistics and Numerical Methods in Chemical Engineering,* Ser. 42. Am. Inst. Chem. Eng., New York, N.Y., 59: 55-63.

Hoskold, H. D., 1889. *Memoire Général et Spécial sur les Mines, la Métallurgie, les Lois sur les Mines, les Resources, les Avantages etc., de l'Exploitation des Mines dans la République Argentine.* Imprimerie et Stéréotypie du "Courrier de la Plata", Buenos Aires, 620 pp.

Hugo, G. R., 1965. How to prepare bids for crown leases sales. *Oil Week,* 16(37).

Kaufman, G. M. 1963. *Statistical Decision and Related Techniques in Oil and Gas Exploration.* Prentice Hall, Englewood Cliffs, N. J., 307 pp.

Kündig, E., 1962. Problems around the evalution and grading of oil prospects. *Bull. Ver. Schweiz. Petrol. Geol. Ingr.,* 29(76): 7-20.

Martinez, A. R., 1966. *Our Gift, our Oil.* Reidel, Dordrecht, 99 pp.

McCrossan, R. G., 1968. An analysis of size frequency distributions of oil and gas reserves of western Canada. In: GEOLOGICAL SURVEY OF CANADA, DEPARTMENT OF ENERGY, MINES AND RESOURCES (Editor), *Report of Activities, Part B, Nov. 1967—March 1968.* Roger Duhamel, Ottawa, Ont.

Merret, A. J. and Sykes, 1962. *The Finance and Analysis of Capital Projects.* Longmans Green, London, 544 pp.

Mood, A. M. and Graybill, F. A., 1963. *Introduction to the Theory of Statistics.* Mc.Graw Hill, New York, N.Y., 443 pp.

Rogers, L. C., 1969. Independents jump on the drilling "fund" bandwagon. *Oil Gas J.,* March 31: p. 31-33.

Schoemaker, R. P., 1963. A graphical short-cut for rate of return determinations. *World Oil,* 157(1): 72-84; 157(2): 69-73; 157(4): 64-68.

Timm, B. C., 1969. Use computers to boost production at a profit. *World Oil,* 168(1): 60-69.

Van der Laan, G., 1968. Physical properties of the reservoir and volume of gas initially in place. IN: KONINKLIJK NEDERLANDS GEOLOGISCH EN MIJNBOUWKUNDIG GENOOTSCHAP (Editor), *Symposion on the Groningen Gas Field.—Verhandel Koninkl. Ned. Geol. Mijnbouwk. Genoot., Geol. Ser.,* 25: 25-33.

Weeks, L. G., 1965. World offshore petroleum resources. *Bull. Am. Assoc. Petrol. Geologists.,* 49(10): 1680-1693.

Wooddy, L. D. jr. and Capshaw, T. D., 1960. Investment evaluation by present value profile. In: SOCIETY OF PETROLEUM ENGINEERS OF A.I.M.E. (Editor), *Oil and Gas Property Evaluation and Reserve Estimates.* Petrol. Trans. Reprint Ser. 3: 171-174.

Chapter V

The influence of mining legislation; standpoint of the company

Introduction

The purpose of this chapter is to study the influence of mining legislation on the profitability of a project, from a company's point of view.

Each piece of mining legislation has its own peculiar aspects. Some regulations require the payment of a bonus for the concession, a royalty on the production and a corporate-income tax from which a depletion allowance may be deducted. In other cases royalties must be paid according to a sliding scale; there is an income tax but no depletion allowance; and participation of the state is required to a certain degree. These different types of legislation influence the profitability of comparable petroleum-investment projects in distinctly different ways, but how and to what extent?

At the same time identical mining legislation can influence different investment opportunities in vaying ways. Paying a large bonus to obtain a concession may be a very heavy burden for a small project. When large oil fields are expected, however, the same bonus may have little significance in the calculation of the profitability. The question to be investigated is: "What is the influence of various types of mining legislations on the profitability of different investment opportunities?"

First, yardsticks to measure the profitability of the investments must be chosen. This choice is somewhat arbitrary. Three yardsticks have been selected because of their frequent use in the petroleum industry and their theoretical advantages:
(1)   The pay-out time.
(2)   The net-present value.
(3)   The expected-monetary value.

The aim of the analysis in this chapter is not to make exact measurements of the influence of the mining legislation on the profitability of a well-defined project, but to study and evaluate general tendencies.

A piece of mining legislation is a rather complex composite of different components. Each component has its own effect. For instance, what is the influence of a royalty, or the corporate-income tax? Is the influence of the royalty and the corporate-income tax together the sum of the influences of the royalty and the income tax apart? This is definitely not the case, since a royalty can normally be subtracted from the gross income to calculate the taxable-net income. Thus such components must first be examined individually, and only after their solitary effects are clearly understood can their effect in combination be profitably estimated. In this case the parts are the key to unlock the meaning of the whole. Nine different components of a hypothetical mining bill will be studied:
(1)   Initial bonus.
(2)   Bonus at the discovery date.

(3)   Fixed-surface duties.
(4)   Yearly-rising-surface duties.
(5)   Fixed royalties.
(6)   Royalties following a sliding scale.
(7)   Fixed-income tax
(8)   Fixed-income tax with depletion allowance.
(9)   State participation after discovery of a commercial field.

Separate influence of elements on project profitability

To compare these nine different elements, a common level of comparison must exist. It is clear that the influence of each component depends upon the extent to which this component is applied. A royalty of 20% is obviously more important to a company than a royalty of 2%, but how does a royalty of 20% compare with a bonus of U.S.$ 1.— an acre? Such comparisons can be made on the basis of the amount of undiscounted money spent on each of the different components of the legislation. This amount can be spent as a bonus, surface duties, royalties, corporate-income tax, and even as state participation.

Assume that a manager must pay a certain amount of money to the state. The manager can choose to pay this amount in the form of one of the nine different items in a mining bill. What should the relative rating of these nine elements be if they are classified according to their degree of prohibitory effect.

The problem of fixing these relationships is solved in this section by constructing a relative rating according to the pay-out time, the net-present value and the expected-monetary value. The three outcomes will be combined into a single list. Such a list, of course, gives no indication, for instance, whether a manager prefers a royalty to state participation, because the *same* undiscounted amount of money is being compared. In practice the amounts to be paid are widely disparate. Apart from the financial elements of the regulations, other aspects take on added importance. State participation, for instance, means not only the payment of a certain sum to the state, but also partial state-control over operations. The rating list's advantage is that it gives a valuable insight into at least the purely financial aspects of such problems.

Influence on the pay-out time

(1)   The influence of the initial bonus on the pay-out time of a project can be studied in Fig. 11. Curve 1 represents the cumulative-cash flow for this project without the bonus. The pay-out time is given by AB. An initial bonus of the amount of OD shifts curve 1 to curve 2. The initial-pay-out time is now AC' and the project-pay-out time is AC".

(2)   The influence of the bonus at discovery date is exactly the same as that of the initial bonus, because with the cash flow at the discovery date surely negative, the shift of that part of curve 1 after the discovery date is identical to that of the initial bonus.

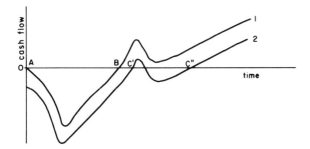

Fig. 11. Influence of bonus on pay-out time

(3) The amount dispensed for fixed-annual-surface duties before the pay-out time has been reached is definitely less than with bonus payments—at least if the pay-out time does not coincide with the end of the lifetime of the project. Thus the influence of the same undiscounted amount on the pay-out time is less marked in the form of fixed-annual-surface duties than as a bonus payment. The fixed-surface duties are accordingly more attractive to the manager than the bonus. The effect of the surface duties can be studied in Fig.12.

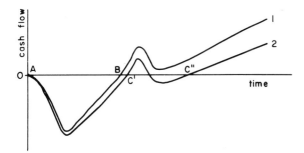

Fig. 12. Influence of surface duties on pay-out time

(4) Yearly-rising-surface duties (but accumulating to the same undiscounted amount as the fixed-surface-duties) show a less marked influence on the pay-out time than fixed-surface duties in all comparable situations. A greater share of the duties in this case must be paid after the pay-out time has been reached.

(5) If the same amount of money is paid in the form of a fixed royalty, its effects upon pay-out time differ from that of surface duties. Surface duties must be paid from the date the licence has been granted and royalties from the start of production. On the other hand, production per well may be higher in the first years of the project, and more royalties must be paid accordingly. Fixed royalties can be regarded on the average as slightly preferable for the manager. The influence of such royalties can be studied in Fig.13.

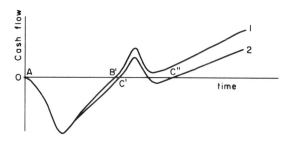

Fig. 13. Influence of royalty on pay-out time

(6)   The royalty according to a sliding scale is normally such that a low royalty adheres
      to a small production and a high royalty to a large production. This means that
      virtually no influence of this royalty can be expected on the pay-out time of a small
      field, and a clear influence on a large field.
      For a large but low-profit field (for example, a field with a low productivity per
      well) the influence of this type royalty is severe. A high royalty is demanded by the
      extensive production; but with minimal profits even before its application, the
      royalty cuts heavily into this small gain, with a fair possibility of making such a
      field sub-marginal. The comparison between a fixed royalty and a sliding-scale
      royalty is not possible on the basis of the same amount of money if the production
      scheme is not known. It is, therefore, better to leave the rating of this element for
      the last part of this section.

(7)   The consequence of the corporate-income tax for the pay-out time contrasts
      sharply with the royalty. Corporate-income tax is normally applied as a percentage
      of the taxable-net income; and the taxable-net income is generated only at a time
      when cash flow according to the tax law is positive. This point in time is usually not
      too far removed from the pay-out time. The tax payments begin shortly before the
      pay-out time is reached because a large part of the outlays can be recovered only
      through depreciation. Consequently the influence of the corporate-income tax on
      the pay-out time is limited as can be seen in Fig. 14. Therefore the corporate-income
      tax must be rated more attractive than the royalties to a manager.

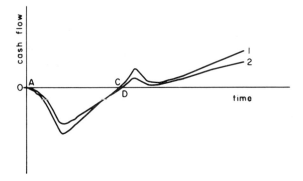

Fig. 14. Influence of corporate income tax on pay-out time

(8)   When a depletion allowance is subtracted from the calculation of taxable-net income the influence on the pay-out time remains nearly the same, for there is ultimately a comparable amount of money extracted on an undiscounted basis (as was the assumption in this section).

(9)   State participation after a discovery influences the pay-out time not at all, if the state can be charged in an orderly way for its share of the outlays following its decision to participate. For instance, a state participation of 50% simply diminishes the cash flow by 50%. This holds for both negative and positive cash flows. An illustration is given in Fig. 15.

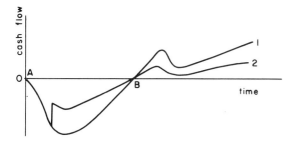

Fig. 15. Influence of state participation on pay-out time

If a manager uses the pay-out time as a yardstick to rate the different elements of a possible piece of mining legislation according to their detrimental effects, the rating, beginning with the least attractive method of payment to the state, would be as follows: initial bonus and bonus at the discovery date, fixed-annual-surface duties and fixed royalties, rising annual-surface duties, corporate-income tax with and without depletion allowance and state participation. The sliding-scale royalties cannot be properly rated.

Influence on the net-present value

The interest rate is an important factor in the calculation of the net-present value. The higher this rate the lower the net-present value. Therefore, all components must be compared on the basis of the same interest rate, although its exact value is not significant for the comparison. It can be assumed that this rate lies somewhere between 8-12%. The influence of the different regulations on the net-present value is highly dependent upon the temporal pattern of the payments. The later the payment to the state, the less the influence on the net-present value.

(1)   An initial bonus is a heavy negative weight upon the net-present value of a project. Such a bonus must be accounted at its full value because there is no discount on this outlay. The later the bonus is paid the less its present value

(2)   Thus, a bonus paid at the discovery date has a smaller present value than an initial bonus, and its negative weight is not so great.

(3)   For an equivalent amount paid in fixed-surface duties the present value is less than the initial bonus, and generally also less than a bonus paid at a later date (the bonus should be paid during the first half of the lifetime of the project to obtain a higher net-present value than the fixed-surface rights).

(4)   From the standpoint of the manager the situation is better with rising-surface duties because a greater amount of such duties is paid against the end of the life of the project.

(5)   Fixed royalties have approximately the same influence as fixed-surface duties. The payment of royalties begins later than the payment of the surface duties, but important amounts are often paid immediately following the start of production because of the higher yields in the first period of the lifetime of each well. These two factors generally make the importance of surface duties and royalties (on a comparable undiscounted basis) nearly equal to the net-present value. As soon as yearly production is constant, or nearly constant, however, the royalties must be rated behind the surface duties.

(6)   For the same reason as in the discussion of the pay-out time, the comparison of sliding-scale royalties with the other regulations is not realistic.

(7)   Contrary to the conclusion for the pay-out time the income tax has an important influence on the net-present value. The present value of the income tax depends on the point in time when taxable income has been generated and the type of depreciation used.

         If a certain amount of money must be extracted from the project in the form of income tax, then the present value may be even higher than with royalties because the decline of net income per well may be even more pronounced than the decline of gross income. This means that more income tax than royalties is paid in the beginning of the project on a comparable basis. The amount of taxable income per year naturally depends to a great extent on legal regulations for the calculation of the income tax. The comparison between income tax and royalties is somewhat arbitrary. The best conclusion is that the present value of the income tax may be either higher or lower than a comparable present value in royalties, or surface duties.

(8)   In the U.S.A. the depletion allowance is the permission to subtract 22% of the gross income to calculate the taxable income. This amount should not exceed, however, 50% of the net income before the subtraction of the depletion allowance. For all fields making a reasonable profit, there are almost no yearly depletion allowances falling under the 50% restriction, in their early lifetime. This restriction is important, however, during the last part of the lifetime . When a depletion allowance is permitted, most companies establish their depreciation pattern so that the optimal use can be obtained from this allowance. It was previously noted that the gross income may decline at a slower rate than the net income. If income tax without and with depletion allowance are compared *on the basis of the same undiscounted amount of money* (which means that the rate without depletion allowance is lower than with a depletion allowance) then the present value of the income tax with a depletion allowance may be higher due to the 50% clause. Much depends, however, on the regulations, the production scheme and the handling of taxable income per

field. The effect of the income tax with depletion may result in either a higher or a lower net-present value for the project, compared to the income tax without depletion allowance.

(9)   The effect of the state participation is a fascinating topic. To compare state participation with any of the other components of possible mining legislation, an amount of money paid in the form of state participation must be extracted from the project. Yet, it is unrealistic in this case to speak of a cash payment because state participation consists of partially shared expenses as well as earnings. To make the actual extraction of money comparable to a royalty, for instance, the state's outlays must be subtracted from earnings. The resulting profit can be contrasted with the royalty on an undiscounted basis. With state participation the difference between the interest rate used by the company for the net-present-value calculation and the rate used by the state as a compensation for the company becomes important. This relationship can easily be demonstrated with the present-value profile of a selected example. Table XIII shows three cash flows:

TABLE XIII

FIVE DIFFERENT CASH FLOWS ILLUSTRATING THE EFFECT OF STATE PARTICIPATION AND TAX

No.	0	1	2	3	4	5	6	7	8	. . . . . .	
1.	− 100	−	−	+ 20	+ 20	+ 20	+ 20	+ 20	+ 20	. . . . . .	+ 20
2.	− 100	−	−	+ 70	+ 10	+ 10	+ 10	+ 10	+ 10	. . . . . .	+ 10
3.	− 100	−	−	+ 83	+ 10	+ 10	+ 10	+ 10	+ 10	. . . . . .	+ 10
4.	− 100	+ 10	+ 10	+ 20	+ 20	+ 20	+ 10	+ 10	+ 10	. . . . . .	+ 10
5.	− 50	−	−	+ 10	+ 10	+ 10	+ 10	+ 10	+ 10	. . . . . .	+ 10

Cash flow 1 is the original cash flow: U.S.$ 100.− investment, U.S.$ 20.− earned at the end of each year, starting at the end of year 3.
Cash flow 2 is cash flow 1 with state participation of 50%, and compensation at an interest rate of 6%.
Cash flow 3 is cash flow 1 with state participation of 50%, and compensation according to the internal rate-of-return of cash flow 1.
Cash flow 4 is cash flow 1 with a tax of 50%, and with a straight line depreciation over 5 years for the invested capital.
Cash flow 5 is cash flow 1 with a tax of 50%, and with a possibility for direct write-off of the tax credit.

The net-present-value profiles of the five cash flows are given in Fig. 16.

(a)   The first cash flow represents a project without government participation. After an investment of U.S.$ 100.− at the beginning of the project, an infinite series of earnings commences at the end of year 3. These earnings are U.S.$ 20.− at the end of each year. The "internal-rate-of-return" of this project is 13.6%;

(b)   The second cash flow includes state participation of 50%, after the discovery of a commercial field. Again, the company invests U.S.$ 100.− and at the end of year 3 begins collecting a series of earnings of U.S.$ 10.— per year. But at the end of year 3 the state compensates the private company's outlays by paying U.S.$ 50.— plus a compound interest of 6% over three years, amounting to roughly U.S.$ 10.—. This brings the company's earnings at the finish of year 3 to U.S.$ 70.—. The "internal rate-of-return" of this cash flow is only 12.6% as can be seen in Fig. 16, Curve 2;

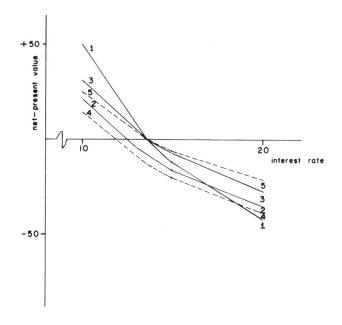

Fig. 16. State participation and the interest rate applied for compensation. The curves illustrate cash flows that are given in Table XIII

(c) The third cash flow demonstrates that the "internal-rate-of-return" would not have changed if the government had paid a compound interest of 13.6%, or the original rate of return, for compensation to the company after the three year's time lapse. The present-value profile resulting from such government participation is shown in Curve 3.

Note how the net-present value of a project calculated at a rate of 10% is influenced by state participation. The net-present value of the project *with* state participation (Curve 2 and 3) is less than the net-present value of the project *without* state participation. At a rate of 15% it is profitable to accept state participation because it can be assumed that the state's compensation can be reinvested at 15%! Clearly the effect of state participation fluctuates according to the interest rate used for calculations and the interest rate applied by the state for compensation.

A comparison with a comparable amount of corporate-income tax is made in Curve 4 in Fig. 16. This curve shows the present-value profile of the project as in Curve 1, with the addition of a corporate-income tax. The investment of U.S.$ 100.— can be recovered in five years by a straight-line depreciation. Comparison of Curve 3 and 4 reveals that state participation is more profitable than a corporate-income tax at these interest rates. However, much depends upon the tax law. If all outlays can be treated as direct write-offs for tax purposes, then Curve 5 is the result; in such a situation the corporate-income tax is preferable to the company, at least when the two cases are compared at an interest rate higher than that which state wishes to apply for compensation. In general, outlays must be partially capitalized; thus in our example the actual curve would fall between 4 and 5 or at approximately Curve 3. Although much depends upon income-tax regulations, the different interest rates applied by company and state, and the actual cash flow, it can be concluded that government participation influences the net-present value of a project in much the same way that the corporate-income tax does.

If a manager uses the net-present value as a yardstick, the relative rating would be as follows: initial bonus; bonus at the discovery date; then a group of elements comprising: fixed royalties, income tax with and without depletion allowance and state participation; and finally, rising-surface rights. The sliding-scale royalties cannot be classified. It is striking that a large group of elements has a markedly similar effect on the net-present value of the project when compared on the same undiscounted basis.

Influence on the expected-monetary value

As illustrated in Chapter IV, the expected-monetary value depends heavily upon the probability-of-success, as well as the distribution of outlays between the pre- and post-discovery periods. Some components of mining legislation adhere to pre-discovery outlays, like the initial bonus; other regulations accompany post-discovery outlays, like the royalty. This factor is an important one in the evaluation of the different components.

Mining legislation's influence on the expected-monetary value is illustrated with an example presented in Table XIV. The table shows that an investment of U.S.$ 100.— is followed by a series of proceeds of U.S.$ 100.— a year beginning in the middle of the third year. The probability-of-success is 40% and the interest rate 10%.

(1) An initial bonus must be regarded entirely as a pre-discovery outlay. This implies that the value of this bonus must be subtracted from the original expected-monetary value of the project to obtain the new expected-monetary value. The expected-monetary value, based on mid-year revenues discounted at 10%, and with a probability-of-success of 40%, is in our example U.S.$ 103.—. After the subtraction of the bonus of U.S.$ 100.— the new expected-monetary value is U.S.$ 3.—.

(2) Two components of mining legislation can be regarded as figuring partly in pre-discovery and partly in post-discovery outlays: fixed-annual-surface duties and rising-surface duties. The expected-monetary value of an undiscounted amount of U.S.$ 100.— diminishes less with rising-surface duties than with fixed ones; this is because a smaller part of the former can be regarded as pre-discovery outlays. Table XIV has shown that fixed-surface duties result in an expected-monetary value of U.S.$ 71.— and rising-surface duties in an expected-monetary value of U.S.$ 74.—.

TABLE XIV

INFLUENCE OF MINING LEGISLATION ON THE EXPECTED-MONETARY VALUE FOR A SELECTED EXAMPLE

Note: probability-of-success is 0.4
       interest rate is 10%

type of cash flow	time pattern							expected-monetary value
	year 0	year 1	year 2	year 3	year 4	year	year 12	
original cash flow	− 100	---	---	+ 100	+ 100	....	+ 100	+ 103
initial bonus of − 100	− 200	---	---	+ 100	+ 100	....	+ 100	+ 3
bonus at discovery date − 100	− 100	---	---	0	+ 100	....	+ 100	+ 71
fixed-annual-surface duties	− 100	− 8.3	− 8.3	+ 91.7	+ 91.7	....	+ 91.7	+ 71
rising-annual-surface duties	− 100	− 5	− 5	+ 91	+ 91	....	+ 91	+ 74
royalty, accumulating to − 100	− 100	---	---	+ 90	+ 90	....	+ 90	+ 83
tax rate 11.1% 50% direct write-off 50% straight line	− 94.4	+ 0.46	+ 0.46	+ 89.45	+ 89.45	....	+ 89.45	+ 89
state participation 11.1% 6% interest for compensation	− 100	---	---	+ 101.9	+ 88.9	+ 111	+ 88.9	+ 86

(3)   All the other possible components are extracted from the project after the discovery. Of this group, the most severe influence is wielded by the bonus at discovery date, as demonstrated by an expected-monetary value of U.S.$ 71.—. The present value of this bonus is unusually high because this outlay is required immediately after the discovery and in a lump sum, as can be seen by cash flow in Table XIV.

In contrast, the royalty spreads the same amount over the production life of the project, consequently lowering the amount that needs to be subtracted from the expected-monetary value. The resulting expected-monetary value is U.S.$ 83.—.

The influence of the corporate-income tax is, as is previously noted, dependent on the income-tax regulations. An arbitrary depreciation scheme has been chosen. The expected-monetary value is in this case U.S.$ 89.—.

These influences are calculated on the assumption that a tax credit can be obtained by crediting the outlays against company profits from other projects. If it is not possible, or if income-tax regulations forbid subtraction of exploration costs that do not result in production, then the expected-monetary value falls considerably, to U.S.$ 84.—. In general, however, a tax credit for outlays can be partially or wholly obtained.

The effect of a possible depletion allowance cannot be illustrated in this example, because when compared to the same disbursment of U.S.$ 100.— to the state there is no difference evident.

The influence of state participation is more negative than the hypothesized income-tax regulation, at least at the 10% interest rate. State participation begins only when the discovery stage has been reached. The "participation-credit" of U.S.$ 12.90 can only be obtained when a commercial field has been found. There is a chance of 60%, however, that no "participation-credit" can be obtained, following the setback of a dry hole. If, however, the pre-discovery investment of U.S.$ 100.— cannot be deducted for the calculation of taxable-net income, state participation is better for the company—the 11.8% tax rate is wholly comparable to the 11.1% rate of state participation, except that the tax credit must be partially recovered by depreciation, while the "participation-credit" can be obtained in the middle of the third year.

The example given in Table XIV is somewhat arbitrary. Important is the probability-of-success. The lower the probability-of-success the less the relative weight of the post-discovery outlays with regard to the pre-discovery outlays.

The components of the mining legislation have been divided into three groups:
(a)   component belonging to the pre-discovery outlays: initial bonus;
(b)   components belonging partly to the pre-discovery outlays and partly to the post-discovery outlays: surface duties;
(c)   components belonging to the post-discovery investments: bonus at the discovery date, royalties, income tax and state participation.

The contrasts in influence among these three groups becomes progressively more discovery outlays: surface duties;

Another inadequacy of this example, however, is that there is a 40% chance that one particular field will result in one particular cash flow. In practice there emerges an expectation curve of the chance that the profit will exceed a certain value. Here the different components cannot possibly be compared on the same undiscounted basis, because the expectation curve deals with an expected range of profits. The negative influence of the pre-discovery extra outlays is static but the influence of the post-discovery outlays changes.

The different influences of each component are almost unpredictable because the expectation curve may have any shape. The analysis of the three previously mentioned groups, however, remains valid. At a comparable investment level, the influence of these groups falls in a roughly descending order:
        pre-discovery components;
        pre- and post-discovery components;
        post-discovery components;
and their differences become more clear as the probability-of-success diminishes. That this ranking is far from refined, can be concluded immediately from Table XIV, where a post-discovery bonus has a more pronounced influence than rising-surface duties.

Sliding-scale royalty

The influence of the sliding-scale royalty remains to be described. In Fig. 17 an expectation curve for a selected prospect has been drawn. After the application of a fixed royalty, the expectation of the net-present value diminishes according to the horizontally hatched area, resulting in a new expectation curve for the prospect inclusive of a fixed royalty. A sliding-scale royalty generally results in a "steeper" expectation curve than a fixed royalty, because the large-profit fields are likely to require a larger royalty than the small-profit fields. The original prospect-expectation curve must now be diminished by the vertically hatched area. The form of the horizontally and vertically hatched areas is fixed by the form of the original expectation curve, the distribution of the expected-production figures, the sliding scale applied, and the rate for the fixed royalties.

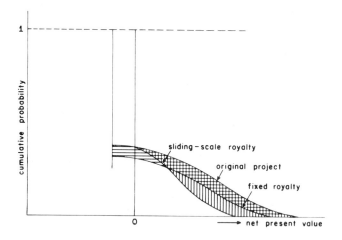

Fig. 17. Influence of sliding-scale royalties on the expectation curve

The fixed royalty, as opposed to the sliding-scale royalty, tends to have a more negative influence on the probability-of-success. If the sliding scale and the distribution of the expected fields is such that the royalty paid for the marginal fields is zero, then the probability-of-success is not influenced by this royalty. This is also the reasoning behind the sliding-scale royalty: making more fields intra-marginal. This positive aspect is for the private companies offset by the negative aspect of less gain on profitable fields. On the "average" the economic effects of the sliding-scale royalty and the fixed royalties tend to converge.

Conclusions

From the foregoing discussions in this section we can conclude that the various components of a mining legislation influence the degree of profitability of investment projects differently. This influence is highly dependent on the probability-of-success—or the degree of risk involved—and the cash flow of each project. It is instructive, however, to summarize the results of the previous discussions into a single list. This would give a general idea of how the average manager would evaluate the impact of each different element. In the order of descending negative influence on the profitability of the project, the components can be listed as follows:

(1)   Initial bonus—This component has the maximum influence on the pay-out time, the net-present value and the expected-monetary value; it, therefore, heads the list.

(2)   Bonus at discovery date—This component has the maximum influence on the pay-out time, is ranked second for the net-present value, and weighs heavily on the expected-monetary value. This last influence can only be ranked second if the probability-of-success is not too low and the discovery date not too late.

(3)   Fixed-surface duties—This item is ranked third for the pay-out time, with others is ranked third for the net-present value, and is ranked second or third for the expected-monetary value.

(4)   Rising-surface duties—The proper place of this component is debatable. It comes after the fixed royalty in effect upon the pay-out time, and at the bottom of the net-present value list; yet it is partially a pre-discovery investment.

(5)   Fixed and sliding-scale royalties—The fixed royalty comes clearly before income tax and state participation on the pay-out-time list. The difference with regard to the net-present value and the expected-monetary value, however, is not important. The sliding-scale royalty is grouped with fixed royalties for previously described reasons.

(6)   State participation—State participation follows slightly behind income tax in the pay-out time calculations, shows no distinguishable difference from income tax in its effect on the net-present value, but influences the expected-monetary value more heavily than income tax (if it is assumed that exploration investments can be deducted for the calculation of the taxable income). The lower the probability-of-success the greater this difference. For this reason state participation has been ranked before income tax.

(7)   Corporate-income tax—There are no predictable differences between income tax with and without depletion allowance (of course compared on the same undiscounted basis); thus these two groups can reasonably share the last place on the list.

Influence of mining legislation on the profitability of a project

The different components of mining legislation partially influence each other. As a result the combined influence of these elements does not simply amount to the summation of their separate influences. A key factor in these interrelationships is the corporate-income-tax regulation. It is assumed in the following discussion that the bonus, royalties and surface duties can be treated as direct write-offs. If state participation is included in the mining legislation, it is assumed that all previous outlays are included in the calculation for compensation. This means that the bonus is also partially compensated by the state at a later date. These assumptions are not always consistent with existing petroleum legislation, but this does not change the general conclusions that will be drawn from this analysis.

The bonus, surface duties and royalties show no direct influence on each other. An indirect relationship, however, can exist between annual-surface duties and royalties. If the royalties or surface duties are high, marginal production toward the end of the project may be cut off. This causes an adjustment in the ultimate production and lifetime of the project, which in turn alters the royalties and the surface duties to be paid.

The bonus, surface duties and royalties determine rather strictly the amount of income tax, being a fixed percentage of the taxable income. The same three components plus the income tax determine the amount that can be expected to be lost to state participation.

Influence on the pay-out time

From the previous section it has become clear that the effect of state participation and the corporate-income tax on the pay-out time is nil or minimal. It follows that only the bonus, surface duties and royalties influence the pay-out time.

The combined influence of bonus, surface duties and royalties results in a more protracted pay-out time than would be expected from their individual influences. Since surface duties and royalties must be paid at regular time intervals, these public revenues grow in value the more the pay-out time is delayed, as shown in Fig. 18.

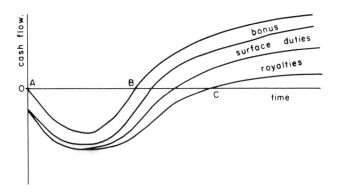

Fig. 18. The cumulative influence of bonus, surface duties, and royalties on the pay-out time

Influence on the net-present value

Before the influence of mining legislation on the net-present value of a project is outlined, it is necessary to examine in greater detail the method for the calculation of corporate-income tax. The description of BIERMAN and SMIDT (1966, p. 122 etc). is useful for this analysis. It is assumed that the investments have no salvage value at the end of the lifetime of the project. This generally holds true for the majority of investments in petroleum exploration and production because the salvage value is nil or very small, especially when its present value is considered. The proceeds generating from an investment without salvage value can be calculated as follows:

before-tax cash proceeds = revenues − cash outlays.

In this equation it is assumed that all revenues are accompanied by an immediate generation of cash equal to the revenues, and cash outlays are equal to the expenses (excluding depreciation). The tax law may give certain rules for depreciation, or it may include a depletion allowance, etc. In general the after-tax proceeds can be given as:

after-tax proceeds = revenues − cash outlays − income tax

In a number of countries the income tax is calculated as a simple fraction of the taxable income, or:

income tax = (tax rate) x (revenues − cash outlays − depreciation) and consequently the after-tax proceeds are:

after-tax proceeds =

(1 − tax rate) x (revenues − cash outlays) + (tax rate) x (depreciation)

In this equation cash outlays include those expenditures required for non-depreciable assets as well as other disbursements charged to current expenses. With the calculation of the present value of the after-tax proceeds, the method of depreciation of the depreciable assets becomes important.

BIERMAN and SMIDT (1966) discuss three methods of depreciation: the straight-line method, the sum-of-the-years'-digits method and the declining-balance method. In addition the unit-of-production method is popular in evaluating oil and gas projects. This method depreciates the asset with amounts equivalent to the yearly production. Different methods are understandably selected by the company according to their profitability per specific project. A high initial depreciation is often desirable due to the high present value of the amounts that can be subtracted from taxable income; yet the tax structure sometimes dictates the use of a more moderate depreciation scheme. For instance, when a depletion allowance is included in the tax structure, the positive effect of a high depreciation rate in the early years of the project can be offset by the lower depletion allowance (as percentage of the net income) obtained.

The net-present value of any project can be calculated as: NPV = PV after-tax proceeds − PV Investment.

The combined effect of the different elements of a piece of mining legislation can be given in a single formula, if bonus, surface duties, and royalties are regarded as the outlays to be subtracted from revenues in calculating the taxable income. This formula is:

NPV = (1-s) x { (1-t) x (PV revenues − PV bonus − PV surface duties − PV royalties − PV cash outlays) + (t) x (PV depreciation) − PV Investments }− S

In this formula s represents the degree of state participation expressed as a fraction between 0 and 1, and t represents the tax rate; while S is a figure that gives the loss due to the difference in interest rates applied by the state and the company during the compensations of the state to the company.

The combined influence of the components follows clearly from the formula. The effect of a rise in bonus, surface rights or royalties will be softened by the effect of corporate-income tax. State participation lessens the influence of all other components of mining legislation.

Contrary to what results with pay-out time, the influence of bonus, surface rights and royalties on the net-present value is the sum of their separate influence. At a certain interest rate, the amount S can be such that an intra-marginal project can become sub-marginal due to state participation. The same is true for the income tax, since PV Investment $-$ (t) x PV Depreciation is more than $(1-t)$ x PV Investment.

Influence on the expected-monetary value

The influence of the combination of the different components of mining legislation on the expected-monetary value is to a great extent determined by the pre- or post-discovery character of the various components.

The expected-monetary value can be obtained most simply by comparing the weighted averages of all possible net-present-value outcomes (after tax and other pay-ments to the government) with the pre-discovery outlays and investments. This can be illustrated with the following formula:

$$\text{EMV} = p \text{ x expected NPV} - q \left\{ \text{ x } (1\text{-}t) \text{ x (PV pre-disc.} - \text{Public Revenues) + PV} \right.$$
$$\left. \text{pre-disc. Investment} - (t) \text{ (PV pre-disc. Depreciation) } \right\}$$

In this formula:

p is the probability-of-success,

q is the probability of a dry hole, or $q = 1 - p$

and the expected NPV is the weighted average of all the possible outcomes for the net-present value, including those in which state participation is calculated, after tax and other payments to the government. The pre-discovery-Public Revenues are the initial bonus and the pre-discovery-surface duties. It is not always possible to claim pre-disco-very depreciation as a deduction from taxable income. The tax law must permit deduc-tion of exploration losses.

In principle, the effect of the combination of elements is very similar to the effect on the net-present value. State participation and corporate-income tax diminish the effect of bonus, surface duties or royalties, and the combination of the influence of bonus, surface rights and royalties is the sum of their separate influences. The weight, however, of the different components varies widely. For instance, the extent to wich state parti-cipation diminishes the influence of the bonus and the surface duties depends entirely on the value of q.

Conclusions

The components of mining legislation are listed below in the order of their dimini-shing negative influence: initial bonus, bonus at discovery date, fixed-surface duties, rising-surface duties, fixed royalties and sliding-scale royalties, state participation and corporate-income tax. The combination of the different components affects differently pay-out time on the one hand and the net-present value and expected-monetary value on the other. Bonus, surface rights and royalties have a greater negative influence on the pay-out time then the sum of their separate influence would suggest; while state participa-tion and corporate-income tax have no significant influence. These last two elements, however, are important for their effect upon net-present value and expected-monetary value—besides diminishing these two values directly, they may also decrease indirectly the amount that must be paid as bonus, surface duties or royalties. All the elements, seperately or in combination, are able to convert an intra-marginal project into a sub-marginal one.

It generally emerges that mining legislation which emphasizes the bonus, surface duties and royalties is less attractive to a company than legislation with accents on corporate-income tax and state participation, at least from a purely financial standpoint. For this reason, in a large proportion of mining legislation the amount of money that must be paid in the form of bonus, surface duties or royalties is less than the amount lost in the form of corporate-income tax or state participation. This leads us to the following chapter where the influence of mining legislation will be considered from a government standpoint.

Literature

Bierman, H. and Smidt, S., 1966. *The Capital Budgeting Decision. Economic Analysis and Financing of Investment Projects.* McMillan, New York, N.Y., 420 pp.

Chapter VI

Influence of mining legislation; the government standpoint

In this thesis petroleum legislation's influence on solely the exploration and production of oil and gas will be studied. Such a study can be extended to cover transport, refining, and marketing of products but these topics are beyond the scope of this effort. In any government policy, however, these "downstream operations" must be encompassed by the government standpoint, because these activities may contribute considerably to the welfare and development of a country.

Introduction

The final shape of the mining legislation of a given country is determined by numerous factors. It depends heavily upon the mineral policy of its government, which in turn rests upon the political and economic situation.

Important considerations with regard to the mineral policy are: the degree of unemployment in a particular country, the need for increased foreign exchange, the impact which possible mining projects may have on other sectors of the nation's economy, such as transport or energy production, and the possibility of reducing imports. Other influential factors are the historical development of the petroleum policy, the interests of various groups in the development of the petroleum industry (as domestic petroleum producers, transport companies, chemical companies, other energy producers, etc.) and the economic-geographical setting of the country.

Mining legislation's target is to promote so far as possible positive benefits accruing to the society within the state from the exploration and production of the minerals. Government is obligated to regulate mining activities in such a way that an optimum income to the state and the society will issue from these operations. All the previously mentioned factors—unemployment, the need for foreign exchange, etc.—will contribute to the establishment of the level of the *optimum income*. Specific definition of this term is a matter assigned to economists. An example of the definition of such an optimum income is given by WOLFF (1964) (cf. Chapter III). Wolff defined the national economic value of a deposit as the sum of the private company's profit, the earnings by the government and the consumer's surplus.

One of the above variables, the level of direct earnings derived by the state, has significant geological aspects. The direct earnings to the state are a key component of each mining or petroleum law. Particularly in small, or developing countries, where the greater part of the produced petroleum is exported, and large percentages of company profits are transferred to other countries, the optimum income will be determined by the direct government earnings.

Another important economic element, omitted from the following discussion must be mentioned in passing: the multiplier effect of the reinvestment of the company profits, and other commercial outlays, as well as the multiplier effect of the government outlays made possible by the earnings from petroleum exploration and production. This effect can be considerable.

To establish the level of the optimum income to the state, insight must be acquired into the maximum public revenue possible from the mining activities. The *maximum public revenue* will be defined as the largest amount of revenues (expressed in present value units) that can be earned directly through taxes and other payments due to the state from independent mining companies operating within the boundaries of such a state. The maximum public revenue can be earned in the form of bonuses, surface duties, royalties, corporate-income taxes, state participation, or various combinations of these methods.

The general outline of the range of choice involved in deriving this maximum public revenue is simple. If the government requires no payments, the exploration and production conditions will be highly favorable (at least as far as mineral legislation is concerned), and an extensive effort on the company's part will result. If the government raises payments which are too burdensome, there will be no exploration or production because the companies find projects unattractive. In these two extreme cases, no income at all will accrue to the state, in the first case because no payments were required and, in the second case, because not a single company was willing to explore or produce petroleum. The maximum public revenue clearly lies somewhere between these two extremes.

A key-factor in this study is the manner in which companies react to proposed mining legislation. The determination of whether a project is attractive or not lies with the private or state-owned company. The yardsticks for the determination of the profitability were discussed in the previous chapter, where the pay-out time, the net-present value and the expected-monetary value were used. The pay-out time is not a useful yardstick for the purpose of the present chapter. The various components of a particular piece of mining legislation do not have a clearly demonstrable influence on the pay-out time. The amount of influence depends heavily on the cash flow of the project. The effect on the net-present value and the expected-monetary value is established by both the cash flow of the element and the pre- or post-discovery character of the payment. Thus the quantitative influence of each change in mining legislation on the net-present value or expected-monetary value is rendered predictable. For the discussion of the reaction of the companies to government proposals, the concepts of the net-present value and the expected-monetary value seem useful. But it must be remembered that the conclusions are only valid against this background.

The concept of the conditions pressure

Normally the geographical area covered by mining legislation is large enough to contain several projects of interest. These projects will differ in worth, and mining legislation will naturally affect them differently.

The influence exerted by petroleum law will be termed the *conditions pressure*. By raising one or more types of payments the conditions pressure becomes higher and by lowering these payments the pressure becomes lower. It becomes more difficult to predict a change in conditions pressure when a petroleum law raises payments for one component—e.g., royalties—and simultaneously lowers payments for another—e.g., the bonus. Such a change in petroleum legislation may even result in the same conditions pressure.

As stated previously, the concepts to be manipulated in this chapter are the net-present value and the expected-monetary value. It is therefore useful to define conditions pressure according to these two diffirent yardsticks. The conditions pressure in terms of the net-present value can be defined with the following formula:

$$CP = \frac{NPV - NPV'}{NPV}$$

in which NPV is the net-present value of the project if no payments to the government are necessary; and NPV' is the net-present value if such payments are required.

If no payments are received by the government, NPV' is equal to NPV and the conditions pressure is zero. When CP is 1, the net-present value after payments to the government is zero, and the project is marginal for the producer. If the NPV' becomes negative the conditions pressure becomes greater than 1. In this case the fields are submarginal from the standpoint of the producer.

It must be emphasized that the yardstick of the net-present value is important for theoretical purposes only. It can be used only if the cash proceeds and the cash outlays are known. Since uncertainty is a fact of life in the petroleum industry, it is impossible to obtain the required certainty for the application of the net-present-value method to important investment problems such as the payment of a bonus for a concession or the drilling of an exploration well.

Any calculation of the conditions pressure is therefore necessarily of theoretical value only, since the conditions pressure is calculated to the exclusion of risk. This fact must be clearly recognized lest the temptation overcome an analyst to include the factor risk in the net-present value by using, for instance, replacement costs instead of cash proceeds and outlays attributable to the project. Chapter III explains that replacement costs is a vague concept. An additional way to include risk has been the use of high interest rates, which, as explained in Chapter IV, is undesirable. It is best to maintain the net-present-value concept as it ideally should be: based on a certain cash flow.

Note that by applying this type of net-present value, the rent earned from a certain project is consequently a true one since geological uncertainty is excluded from the concept. A quasi-rent does not exist in theoretical considerations employing the net-present value, because the quasi rent originates from the geological risk. If a petroleum company could produce with certainty from projects now and in the future, no dry holes would be drilled.

It can, therefore, be stated that when the conditions pressure equals 1, the government is earning the entire "true rent" from a hypothetical project. If the government is earning this rent from all the projects within its jurisdictional borders, the maximum public revenue is being obtained according to this theoretical set of conditions.

The advantage of the net-present-value concept lies in the facility with which it promotes a profitable first approach to the complicated problem of the formulation of mining legislation.

The study is of no value, however, if the geological risk is not ultimately included in the analysis; the expected-monetary value yardstick had been chosen for this purpose. The conditions pressure as determined by the expected-monetary value is:

$$CP = \frac{EMV - EMV'}{EMV}$$

The maximum conditions pressure is again 1. In this case a company is barely willing to take the risk to explore. Characteristic of uncertainty is that it changes in degree when increasing amounts of geological information about a specific region become available, or as technical and economic conditions change. This implies that the conditions pressure is changing continuously parallel to the expected-monetary value. The most profound change usually occurs after the results of the exploration well for a project are known. A project that appeared to be barely acceptable to a company may prove to be a failure when a dry hole has been drilled, or the project may prove to be highly successful and the conditions pressure may accordingly be slight.

A crucial period for the government is that moment when the petroleum law applicable to the projects becomes effective. How many companies will accept the terms? How many projects do the companies consider profitable?

When it is assumed that petroleum law is not altered during the lifetime of the projects, the maximum public revenue is provided to the state if the conditions pressure of all the projects was equal to 1 at the date of acceptance of the contracts, with no projects being dropped by the companies due to the terms of the mining legislation.

In this case the maximum public revenue equals only accidentally the true rent of the ultimately successful projects, due to the random element in geological exploration. The companies may be more lucky than they had originally expected, with highly successful projects. The government is confronted with the problem of having requested too little; greater gain could have been made, if more geological information had been available. The government may also be relatively fortunate. The project yields may be rather poor, while the companies have previously paid large bonuses to the government to obtain the concessions. Clearly, the government is not altogether pleased that project yields are poor (directly and indirectly less income for state and inhabitants), thus the preceding use of the qualification "relatively lucky".

It is, however, questionable whether the government is able to fulfil the conditions to reach a maximum public revenue at all.

There generally exists a contrast in the amount of information that is available and capable of being processed by the industry and the government. If companies must place all relevant information at the disposal of the government, and if the concession areas are small, the government may be able to obtain a better insight into the geology than the companies themselves. Normally, however, the companies are better informed, and the conditions pressure can accordingly be calculated more accurately by the companies than by the government. In summary, there exist two sources of governmental uncertainty concerning the value of the conditions pressure: the character of most projects is cloudy, even when the best information is available, and the government is only partially informed of the company's analysis of the project. In a few cases the government is not able to handle all the relevant information thoroughly.

Besides a scarcity of information, the government must contend with difficulties engendered by the structure of the mining legislation itself. Since it can be expected that most projects will vary, the same legislative package may result in considerable differences in the conditions pressure for the various projects. The conditions pressure for a number of projects may be so severe that they become sub-marginal, while conditions pressure for others may be minimal. This results in a considerable loss of earnings for the state, because a number of projects is rendered unprofitable by the mining legislation, while the revenue obtained from others is too small. Mining legislation which maximizes public revenue must surely be *highly selective*. Mining legislation must keep poor projects intra-marginal, and manipulate the very profitable projects to contribute the greater part of the maximum public revenue. Selective mining legislation is the only way to realize the maximum public revenue from mining operations.

Foundations for the analysis in this chapter

Which are the conditions that petroleum legislation must satisfy to attain maximum public revenue?

To isolate these conditions, an analysis in two steps will be carried out. First, a model based on the net-present value will be studied. This model is of solely theoretical value because it excludes the geological risk. Some important general conclusions, however, can be drawn from the model that can be easily modified when the picture is complicated by the introduction of geological risk. This is done in the second model, based on the expected-monetary value. This last model gives the conditions that must be fulfilled for mining legislation to reach the maximum public revenue in an optimal manner.

A model based on the net-present value

To study the influence of mining legislation on different projects using the net-present value scheme, it is first necessary to group the different projects in a logical order. The best way to account for the weight of each project is to study the various projects on a per-barrel basis.

The criteria for grouping different properties can be the net-present value/bbl, the revenues/bbl, the investments/bbl or otherwise. The net-present value/bbl does not seem a logical criterion because it is precisely the value that is influenced by the mining legislation. The revenues/bbl are virtually identical for each prospect because the figure simply represents the present value of the price for each bbl. The investments/bbl form a strong and valuable rating criterion, because the present values of the investments per bbl differ widely for each project and the outlay is not influenced directly by governmental decisions (when all payments to the government are excluded from the term "investment"). Other criteria, such as the cash outlays/bbl, do not appear suitable. This leaves the investment/bbl on a present-value basis (the PV Investment/bbl) as the most reasonable rating criterion.

The PV Investment/bbl can be given in a graph showing the projects grouped according to a rising PV Investment/bbl as is shown in Fig. 19. This figure is of com-

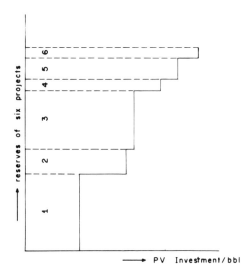

Fig. 19. PV Investment/bbl curve for six selected projects

pletely arbitrary construction, but similar data can be used as a starting point for further deliberation. The next step in such a discussion is to compare the PV Revenues/bbl and the PV Cash Outlays/bbl for the same projects. This gives the net-present values/bbl if no government payments are required. This investigation can be used as a basis for the analysis of the impact of mining legislation, and forms the rough outline of the line of inquiry which will be pursued in this model. The exclusion of all uncertainties implies that the ultimate production—the number of barrels that must be used to calculate the outlays per barrel—is a certain figure. Ultimate production figures are clearly to be preferred to figures for "oil in place" or any other reserve data.

Further, unless otherwise stated, it is assumed that the price is not affected by the number of projects that can be profitably produced. This assumption will not give rise to problems for oil, since the world-oil market is large in comparison with almost all the production from separate fields. The total output may, however, influence the price level in a gas market as is clearly illustrated in Chapter III. In such a case the conclusions of the model must be adapted to the situation.

Another simplification is introduced by assuming that a single product (petroleum) is produced and sold at a single price. Thus no account is made for differences in grades of oil, or for the different prices—on the same energy basis—for oil, natural-gas liquids and natural gas.

PV Investment/bbl curve

The PV Investment/bbl curve must be plotted to fix the different projects in a logical order.

What are the outlays that must be included in "the investment"? For the production of petroleum four types of outlays are important:

(1) Finding investments: outlays for geological and geophysical exploration, drilling and completion of exploratory wells, etc.

(2)   Development investments: equipment costs, development and production wells, secondary-recovery equipment, etc.

(3)   Production investments: the outlays for workovers, which are investments to renew or improve production wells, etc. These investments are generally rather low in relation to the other outlays, and will be omitted from the following discussion.

(4)   Transport investments: the outlays for pipelines or other transport equipment and terminal facilities. Investments for marine shipments or large-distance transport by pipelines will not be included in this model.

All other outlays—apart from payments to the government—can be described as cash outlays and cannot be termed "investment". The types of outlays that must be included in the term "investment" are of course open to discussion; a number of outlays can be interpreted as an investment or as cash outlay, but the actual oil producers generally limit "investment" to those outlays included in the previous list. More elaborate examinations of this subject can be found in HODGES and STEELE (1959) and LOVEJOY et al. (1963).

The following topic then presents itself: what factors influence the shape of the PV Investment/bbl curve? A few pages will be devoted to this question since the shape of this curve is of crucial importance to the determination of the optimal set of conditions that can be established in the petroleum law. The main factors that might influence the PV Investment/bbl are:

(1)   The volume of the ultimate production.

(2)   The productivity of the wells.

(3)   The average depth of the reservoir.

(4)   The geographical position of the project.

(1)   *The volume of the ultimate production*

The volume of the *ultimate production* is a significant factor because it determines the PV Investment/bbl figure directly by establishing the number of barrels available and the amount of the investment. To handle this question in greater detail, investments must be further split into finding, development and transport investments.

(a)   The finding investments are somewhat independent of the volume of the ultimate production. The same expenditure on geological or geophysical exploration may lead to a bonanza or a dry hole. The outlays for exploration may be greater in an attractive area than in an unattractive one, but the relation between exploration investments and the ultimate production of the discovered field is extremely weak. The same is true of the few exploration wells. The larger the ultimate production, the lower the investments/bbl. Consequently, the finding investments/bbl are lower the larger the ultimate production.

(b)   The development investments/bbl are of course heavily influenced by all the four factors previously mentioned—volume of the reserve, productivity per well, average depht of the reservoir and geography. Here we are studying only the relation of the development investments/bbl and the volume of the ultimate production, and we are assuming that the other factors remain equal.

The type of installations necessary to handle a small field offshore varies widely from the types that can be used in handling large or giant ones. The smallest production unit that can be installed offshore is a one-well, one-zone-well jacket. In such a situation, production is through one well that is producing from a single zone. If more producing zones or reservoirs are present in the field, larger well-jackets can be used, with several wells producing from several zones. If large fields are found, production platforms can be used combining a number of wells. Giant fields are produced from several production platforms.

Assuming that all other factors remain unchanged, installation investments/bbl are undoubtedly lower for a multiwell platform than for a single well, and development investments/bbl from a production platform are likely to be lower than from a few scattered well-jackets. Differences in investments per well were illustrated in Chapter III based on the work of CAMERON (1966). On the other hand, when a field is compared with an ultimate production twice as high as another field and requiring twice as many well-jackets or platforms, the development investments/bbl for the two fields are almost the same. The relation between the ultimate production and the development investments/bbl is expected to be irregular. For several ranges of ultimate-production values, the installed equipment is a linear function of this ultimate production (assuming all factors remain equal), as BRADLEY (1967) proposes for a general rule of onshore-development costs. For other ranges, however, there are clearly "economies of scale" with increasing ultimate production. The same is true of secondary recovery projects. For the development investments/bbl offshore it can be concluded that for several ranges of ultimate production these costs will become lower the larger the ultimate production. Onshore this trend is less clear; however, it also exists there. The Slochteren-gas field, for instance, is produced with "well clusters" (groups of deviated wells) that permit considerable "economies of scale" in development drilling costs (BIJL, 1968).

(c)   The transport investments/bbl show considerable "economies of scale" with expected ultimate production. Some examples were given in Chapter III. The larger a field, the larger the possible diameter of the pipelines to terminals. Considerable reduction of transport investments/bbl is found to accompany an increasing field size, because a relative reduction in costs results from the use of larger terminals, larger pipelines, and larger production-transfer facilities.

In summary, an increasing field size generally sees a parallel lowering of the finding investments/bbl, the development investments/bbl and the transport investments/bbl. Consequently, the PV Investment/bbl is extremely sensitive to the ultimate production. This is a clear and strong relationship: the larger the field size the lower the PV Investment/bbl.

(2)   *The productivity of the wells*
When we assume that the ultimate production, the average depth of the reservoir, and

the geographical position remain unchanged, the *productivity of the wells* becomes the parameter under consideration. What is the influence of this productivity on the PV Investment/bbl?

Characteristics of the reservoir, such as the permeability, the porosity, and water saturation determine to a large extent the productivity of the well. The reservoir fluid is considered in this model to be homogeneous, and is called petroleum. This stresses the theoretical character of the model since the properties of oil or gas influence not only the price but also the productivity per well. It is not necessary to go into details of the way and extent to which these factors influence the productivity. Much work has been done on this matter (see, e.g., ARPS et al., 1967).

(a)   The finding investments are virtually independent of the realized productivity per well. This is true for the geological and geophysical work as well as for the outlays for exploratory wells;

(b)   The development investments, however, are highly dependent upon the productivity per well. The greater the productivity per well the lower the development investments/bbl. This is illustrated by the fact that almost the same equipment is employed in production from a well- or badly-producing well. Since the equipment must be renewed after a certain lapse of time, the amount of oil that can be produced from a slowly producing well is less than from a quickly producing one, and the development investments/bbl in the former case are consequently higher. The same holds true for secondary recovery equipment.

(c)   The transport investments are influenced little by the productivity per well. The same amount of oil piped to a terminal, when coming from one well or ten, requires the same amount of investment, involving only some considerations of a few additional facilities in the field.

In considering development investments/bbl it becomes evident that there is a clear trend for PV Investments/bbl to be higher the lower the productivity per well.

(3)   *The average depth of the reservoir*

We will next hold factors such as field size; productivity per well and geographical position constant and examine what effects fluctuations in the third variable, the· average depth of the reservoir, produce.

(a)   The relation of the depth to the costs of geological and geophysical work is manifest as long as very shallow deposits are under consideration. Offshore, a field survey after very shallow deposits is not realistic. For deeper deposits the correlation between costs and depth is weak in these areas. There is a general tendency that exploration becomes more costly when deeper deposits are involved. The costs of exploratory drilling, however, are extremely sensitive to the depth of the expected reservoir. FISHER (1964), as previously mentioned (cf. Chapter III), has pointed out that drilling investments increase sharply with depth. Finding investments/bbl increase likewise when the expected reservoir is deeper in the earth crust;

(b)   The development investments/bbl increase with the depth of the reservoir for the same reasons as the costs for exploratory drilling do—because deeper development and production wells must be drilled to reach the oil-bearing formation. On the surface, heavier equipment must be placed to meet the increased pressures or lifting costs of the petroleum. Secondary-recovery techniques also call for heavier equipment. Especially offshore, however, there are also money-saving aspects of an increasing depth of the reservoir. If possible, more production wells are drilled from one production platform. This is often done with directional drilling to obtain a better spacing of the production points at the level of the reservoir. The deeper the reservoir the larger the possible distance between these production points. Thus the same field that, in a very shallow position, must be produced from individual well-jackets, can in a deeper position be produced from a single platform, which can be more economical. In general, however, the development investments/bbl will increase with greater depth of the reservoir.

The transport investments for oil are influenced little by the depth of the field. For gas, the reservoir pressure can be used for transportation; a deeper gasfield can save on outlays for compression facilities. The effect of the depth on transport investments/bbl is, therefore, uncertain for the assumed homogeneous fluid, petroleum.

A deeper field seems to result in higher finding investments/bbl and higher development investments/bbl, while transport investments/bbl are inconclusive. In general, there will be a tendency for higher PV Investment/bbl as the reservoir lies deeper in the earth's crust.

(4) *The geographical position of the project*

Finally, we will consider the geographical position as a variable. The basic elements of geographical position include the distance from the nearest shore, the depth of the shelf between the location of the field and the shore, the water depth at the location of the field, and the weather in the area in question. Favorable facets of these elements would be when: the position is very near shore, there are no deep throughs between the producing facilities and the shore, the water depth is shallow and the weather is stable and good. Opposite conditions would be classified as unfavorable.

An unfavorable geographical position—assuming all other factors remain constant—leads to higher finding investments. Deep water requires expensive drilling rigs, bad weather interrupts the drilling activities or may cause accidents. The finding investments/bbl are higher as the geographical position becomes less favorable.

Development investments/bbl show a similar relationship. Deep water makes expensive production facilities necessary. The possibility of bad weather forces construction of special equipment to meet all eventualities.

The transport investments/bbl are, of course, markedly sensitive to the geographical location of the project. A location far from the shore, deep water, a deep trough between the production platform and the coast, or bad weather may result in high transport investments/bbl. Similar relations prevail onshore. Therefore, it may be concluded that the PV Investment/bbl is considerably higher as the geographical position becomes less favorable.

In summary the shape of the PV Investment/bbl curve is determined by the four factors in the following way:

(1)   The smaller the expected ultimate production, the higher the PV Investment/bbl.
(2)   The lower the productivity per well, the higher the PV Investment/bbl.
(3)   The deeper the reservoir(s), the higher the PV Investment/bbl.
(4)   The less favorable the geographical position, the higher the PV Investment/bbl.

Next we will control conditions to determine whether there is any correlation among the four factors themselves. At the first glance such a correlation seems absent. For instance, small fields do not necessarily show a low productivity, nor are they automatically deep in the earth crust with an unfavorable geographical position. The four different factors can be arranged into six different combinations that can be checked for their correlation, as is shown in Fig. 20.

There definitely seems to be no indication of a relation existing between the geographical position (water depth, distance from shore, form of seabed, etc.) and the depth of the reservoir, the productivity and the field size. Of course, the topography of a country may be the expression of the underlying geological conditions and it is the task of the geomorphologist to closely study this relationship; however, there is no reason to assume worldwide correlations between the geographical conditions and the depth of the reservoir, the productivity and the ultimate production of the fields. If this were the case, exploration for oil and gas would be considerably simplified.

The amount of oil or gas that has accumulated in the oil- or gas-trap is independent of the depth. Geologists expect that in very deep layers in the earth's crust no oil can exist due to the high temperatures and that at these depths only gas can be expected. There is also some gradual change of the character of the oil, becoming lighter towards deeper formations, but there seems to be no relation between the volume of the ultimate production and the depth of the reservoir.

The properties of the reservoir and the reservoir fluids vary considerably with depth. Since the reservoir fluids, however, are thought to be of a constant homogeneous composition in this model, the conclusions that can be drawn from this variation are not completely valid. Therefore, the correlation is indicated as ± in Fig. 20.

Finally, the last relation shows a positive correlation between the volume of the ultimate production and the productivity per well. The studies of ARPS et al. (1967) proved that the recovery efficiency is higher the larger the oil in place per unit volume of rock. Since the recovery efficiency shows a positive correlation with the productivity per well, it can be assumed that there is a positive correlation between the productivity per well and the expected ultimate production, being the arithmetic product of the oil in place and the recovery factor. Roughly the same conclusion can be drawn from Fig. 21. In this figure the number of bbls/day's production is plotted against the number of producing wells for the producing fields of three regions in the world: Iran, Algeria and northwestern Germany and The Netherlands. It can be seen that the correlation is not particularly strong. The three regions can be represented by rather wide ellipses, but the longest axis of each ellipse is inclining somewhat more than 45°, indicating that the production is rising more strongly than the number of producing wells. Since the production per day shows a strong positive relation to the expected ultimate production of

volume	O	O	+	
productivity	O	±		
depth	O			
geographical position				
	geographical position	depth	productivity	volume

Fig. 20. Relation among the factors: volume, productivity, depth and geographical position

O = no correlation + = positive correlation ± = correlation uncertain

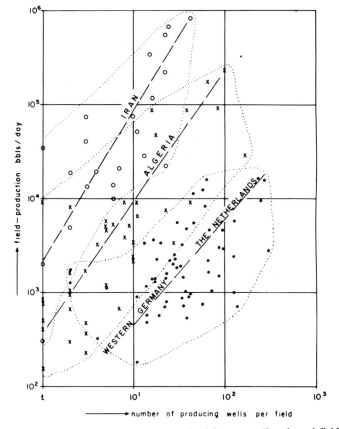

Fig. 21. Relation between productivity per well and total field production, for fields in Iran, Algeria and northwest Europe. Key: *circles* — Iran; *crosses* — Algeria; *dots* — The Netherlands and Western Germany. *Oil Gas J.* 31 December 1969.

the field, it can be stated that the productivity per well will be higher with a larger expected ultimate production. The correlation is weak, however, as can be seen from the width of the ellipses.

The discussion of the various relationship can be concluded by establishing that the only existing relation is a weak, positive one between the expected ultimate production and the productivity. The weak or absent correlations among the factors implies that the shape of the PV Investment/bbl curve can be highly irregular, and that the slope of the curve can vary considerably.

An important factor, however, is the shape of the curve. Can we expect the different PV Investment/bbl values to vary around a straight line? Or will these plotted values form on the average a convex curve (towards the y-axis) or a concave curve?

Information from Chapter II must be drawn upon to answer this question. From Tables II and III it could be concluded that the majority of production issues from only a few giant fields, while a small volume of production comes from numerous small fields. This is an important feature affecting the shape of the PV Investment/bbl curve.

Since the PV Investment/bbl is lower the larger the volume of the expected ultimate production, the largest fields generally can be expected to fall in the lower part of the curve. This is made even more probable by the realization that there exists a weak relation between a large volume and high productivity—resulting as well in a low PV Investment/bbl. This does not exclude the possibility that a large field with a low productivity, far from the shore in very deep formations, might result in a rather high PV Investment/bbl. On the average, however, the larger fields will be represented in the lower part of the curve and the smaller fields in the upper part. This crucially affects the shape of the curve since it means that a large part of the total ultimate production is on the average produced for a rather low PV Investment/bbl. The PV Investment/bbl curve can generally be expected to be convex towards the y-axis.

An idea of how such a curve may look like can be obtained from Fig. 22. ARPS and

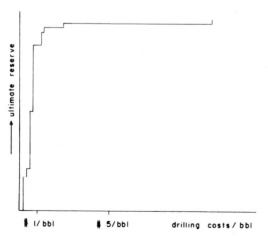

Fig. 22. Drilling costs/bbl curve for the Denver-Julesburg basin
Source: Arps, J. J. and Roberts T. G., 1958. Economics of drilling for Cretaceous oil on the East Flank of the Denver-Julesburg Basin. *Bull. Am. Assoc. Petrol. Geologists*, 42(11):2549-2566.

ROBERTS (1958) studied 338 fields in the Denver-Julesburg basin. In this figure the drilling investments/bbl are given for the different field sizes. From these projects a curve was obtained that was definitely convex towards the y-axis, almost hyperbolic, with the line indicating the ultimate reserve as an asymptote. This discussion on the shape of the PV Investment/bbl curve is based almost entirely on onshore data. It is understandable that offshore the smallest fields are not profitable no matter what the mining legislation may cover or omit. This implies that an offshore PV Investment/bbl curve would be a truncated onshore-type curve.

Another conception of the curve can be obtained if it is realized that in the U.S.A. about 2/3 of the wells are marginal (stripper) wells. These wells account for approximately 1/5 of the total production. If it is assumed that all wells require the same investment, the well costs of 1/5 of the production are about eight times that of the remaining 4/5 part (cf. LEVORSEN, 1967, p. 481).

PV Revenues /bbl curve

The discussion of the PV Revenues/bbl curve can be greatly compressed following the study of the PV Investment/bbl curve. The PV Revenues/bbl is the present value of the expected prices for barrels delivered throughout the life of the project. These prices may change during the lifetime and they may vary from project to project. The price differences between the various projects shall not be too large, since in such case contracts would surely be renegotiated. The present value of the prices of the projects depends, however, on the lifetime of the project and the production scheme. The time for development of the field is usually greater for large fields than for small fields, and the lifetime is also longer. This makes the present value of the total revenues per barrel lower for a large field than for a small one. The revenue is based on realized or transfer prices.

Very long lifetimes occur for the giant fields, for instance, in the Middle East. In this case, however, the fields are only partially developed due to the large production in relation to demand. In this case a giant field must be considered as an object for successive investment opportunities—each additional investment in development being ascribed its own lifetime. These investment opportunities will be identical to those for very profitable fields.

In general, therefore, it can be concluded that the lifetime for marginal fields will be shorter than for large and profitable fields. The PV Revenues/bbl are consequently higher for the marginal fields, and a PV Revenues/bbl curve would bend slightly to the right in the graphs of the marginal fields.

This PV Revenues/bbl curve is completely theoretical, however, because as previously explained, variations in the prices for oil or gas or different types of oil are excluded from the analysis.

PV Operational-Cash Income /bbl curve

When cash outlays are subtracted from revenues, the resulting proceeds can be called the operational-cash income. Cash outlays include only those payments made for the production and transportation of the gas or oil. Exploration and development outlays entirely of an investment character. Cash outlays can be subdivided into:

(a) production costs: outlays for maintaining equipment, regulating the production, etc.;

(b) transport costs: outlays for maintaining transport equipment and regulating transportation to the terminal.

The four factors considered in studying the PV Investments/bbl curve can also be used to study the PV Operational-Cash Income/bbl curve. These factors were: the volume of the possible ultimate production, the productivity of the wells, the average depth of the reservoir, and the geographical position of the project.

(1) The volume of the ultimate production influences the PV Operational-Cash Income/bbl considerably. The cash outlays per bbl are substantially less for a large multiwell platform or a large pipeline than for corresponding smaller ones. The total volume of the cash outlays is greater, of course, but since the amount of the produced oil increases, the cash outlays/bbl diminish. Thus the volume of the ultimate production has much the same influence on the PV Cash Outlays/bbl as it does on the PV Investment/bbl. And the PV Cash Outlays/bbl must be subtracted from the PV Revenues/bbl to obtain the PV Operational-Cash Income/bbl.

(2) The same is true of the productivity per well. The higher this productivity the lower the PV Cash Outlays/bbl.

(3) The depth of the reservoir has various influences, depending on the character of the petroleum.

(4) The geographical position influences the PV Cash Outlays/bbl rather heavily. An unfavorable geographical position results in high cash outlays for production and transport operations. A favorable geographical position results in low cash outlays.

The influence of the four factors on the PV Cash Outlays/bbl is comparable to their influence on the PV Investment/bbl, not unexpectedly. A general rule can be applied that projects requiring low PV Investment/bbl also show a low PV Cash Outlays/bbl, and projects with a high PV Investments/bbl also normally have a high PV Cash Outlays/bbl. The PV Operational-Cash Income/bbl curve is consequently roughly a reflected PV Investment/bbl curve, modified by the effect of the form of the PV Revenues/bbl curve.

The net-present value/bbl

The net-present value per bbl can be studied when the PV Investments/bbl curve and the PV Operational-Cash Income/bbl curve and the PV Operational-Cash Income/bbl curve are given. This is shown in Fig. 23.

This figure shows a situation that might be expected to exist in a selected petroleum producing area. It shows the curves that can be obtained if the projects are arranged according to a rising amount of PV Investment/bbl. The present value is calculated at the interest rate used by the companies to indicate the acceptable internal-rate-of-return. The assumption is made that no risk is involved in the projects and the future revenues are known in advance, enabling the use of the net-present value concept. The interest rate does not take into account the dry-hole risk.

The total number of barrels that can be produced while yielding a positive net-present value for the companies is determined by the point of intersection of the PV Investments/bbl curve and the PV Operational-Cash Income/bbl curve. The economically-recoverable reserve is given on the y-axis as OE. This is the number of barrels that would be produced if the future revenues were completely known and no mining regulations were included in the profitability calculations. An average condition is typified by the Tudor-arch-shaped form of the two curves, as is illustrated in Fig. 23. This form is a key to understanding the conditions that must be applied to maximize the proceeds for the state from mining operations.

The influence of separate components of a petroleum law

The PV Investment/bbl and the PV Cash Outlays/bbl form together the PV Average Costs/bbl. If it is assumed that all the fields are producing at an optimum rate in the company's long-run calculations (in other words, if the field is produced entirely on the basis of the lowest PV Average Costs), the net-present-value/bbl will equal the true rent for each project.

Previously, it was concluded that the maximum public revenue was reached when the entire rent was earned by the government, implying that the conditions pressure for each project should equal 1. The area enclosed by the PV Investment/bbl curve and the PV Operational-Cash Income/bbl curve represents the entire rent on all the projects, or the maximum public revenue, for this model. This will be termed the theoretical-maximum public revenue.

What are the possible methods for obtaining the largest share of this theoretical-maximum public revenue through the various components of petroleum law? The bonus, surface duties, royalties, taxes, and state participation will be discussed to evaluate these possibilities.

(1)   *Bonus*
   The average influence of a given bonus is shown in Fig. 24. Assume that a bonus is required to obtain a concession. This bonus is equal for every surface unit and may be a bonus at the discovery date or an initial bonus. A constant bonus exerts a greater influence on a small field than on a large one. The PV Bonus/bbl is consequently lower for the large fields than for the small. Since the projects in Fig. 24 are arranged according to an increasing PV Investment/bbl and since the larger fields are concentrated on the average in the lower part of the graph, the lower PV Bonus/bbl values can be expected in the lower half of the graph.
   The PV Bonus/bbl is added to the PV Investment/bbl. The consequences of the bonus requirements can be seen from this graph. The economically-recoverable reserve is now OE', as determined by the intersection of the PV Bonus/bbl curve and the PV Operational-Cash Income/bbl curve. The proceeds for the government are only a small part of the theoretical-maximum public revenue. This is because the conditions pressure is low for the larger fields. If the government wants more proceeds in the form of bonuses, the remaining economically-recoverable reserve becomes accordingly smaller. Clearly, the imposition of a large bonus per surface unit is not an efficient means for reaching the maximum public revenue.

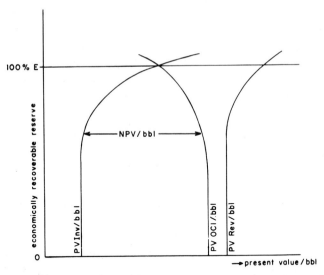

Fig. 23. Tudor-arch-shaped arrangement of PV Investment/bbl curve, PV Operational-Cash Income/bbl curve and PV Revenues/bbl curve

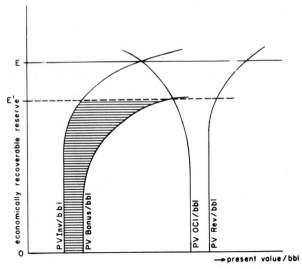

Fig. 24. Influence of the bonus on the economically-recoverable reserve

(2) *Surface duties*

A similar conclusion can be drawn for surface duties. High surface rights cut the marginal fields, while the profitable fields do not contribute proportionally to the government proceeds. The requirement of high surface rights is also an inefficient method for reaching the maximum government proceeds.

(3)   *Royalties*

Somewhat different is the conclusion for the royalties. The *fixed royalty* is expressed as a constant percentage of the price of a unit of petroleum. The proceeds earned by the government are visualized in the hatched area in Fig. 25. The negative effect of the royalties on the economically-recoverable reserve can be expected to be less—on a comparable basis—than the effect of the bonuses and surface rights, because smaller fields contribute less to the government proceeds than larger ones. The royalties are consequently better suited to reach the theoretical-maximum public revenue than are surface duties or bonuses. It is, however, impossible to reach this maximum by requiring royalties, since the economically-recoverable reserve will be lower the higher the royalties that are required.

This problem can be partially solved by using *sliding-scale royalties*. In this case high royalties are required for fields with large yearly production, while low royalties, or no royalties at all, are imposed on the fields with a small yearly production. Since the larger fields appear on the average to be concentrated in the lower part of the graph, high royalties are paid for bonanza's and low royalties on the average for the marginal fields, as is shown in Fig. 26. Since a large Net-Present Value/bbl does not necessarily coincide with a large field (because large, but low-profit fields may exist) the sliding-scale royalty will never be suitable for the complete recovery of theoretical-maximum public revenue. The sliding-scale royalty will achieve this goal, however, better than the fixed royalty.

(4)   *Corporate-income tax*

To study the influence of the corporate-income tax a new curve must be developed: the PV Depreciation/bbl curve. The present values of the depreciation must be known to calculate the present value of the amount of tax that must be paid to the state. The shape of the PV Depreciation/bbl curve will be almost the same as that of

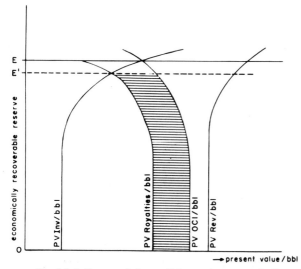

Fig. 25. Influence of the royalties on the economically-recoverable reserve.

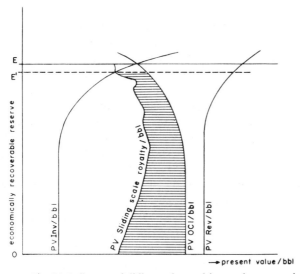

Fig. 26. Influence of sliding-scale royalties on the economically-recoverable reserve

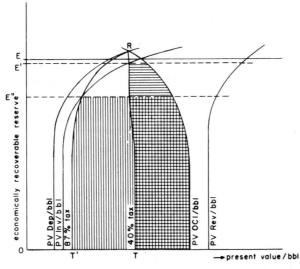

Fig. 27. Influence of corporate-income tax on the economically-recoverable reserve

the PV Investment/bbl curve. The process of depreciation is described in Chapter V. The diminished value of the depreciation with regard to the investment is due to the time lapse between the initiation of the investment and the sequence of successive depreciations of the different assets. The PV Depreciation/bbl curve can, therefore, be obtained from the PV Investment/bbl curve by subtracting a constant part of each PV Investment value in this curve. Exactly what part depends upon the depreciation rules in the tax law.

The amount of tax that must be paid is assigned as a constant percentage of the line segments between the PV Operational-Cash Income/bbl curve and the PV Depreciation/bbl curve—this resulting line segment represents the taxable income. The economically-recoverable reserve is diminished by the requirement of a corporate-income tax from OE of OE'. The line RT is drawn in Fig. 27 in such a way that on the right hand side of this line is given the amount of tax that must be paid. The horizontally-hatched area represents the government proceeds from corporate-income tax. The area between the curve RT and the PV Investment/bbl curve is the net-present value of the remaining projects, or the net-present value after tax.

The higher the tax rate, the farther the curve RT moves to the left. The intersection of the curve RT with the PV Investment/bbl curve determines the economically-recoverable reserve. The farther the curve RT moves towards the left, the lower the value at this point of intersection. As soon as this point reaches the bend of the convex PV Investment/bbl curve, the economically-recoverable reserve diminishes rapidly, as can be seen from the high corporate-income tax resulting in the vertically-hatched area of government proceeds. The angle between the RT' curve and the PV Investment/bbl curve is extremely small. A corporate-income tax that is slightly higher will diminish the economically-recoverable reserve to zero, and will consequently reduce the government proceeds to zero.

It can be concluded from this discussion that corporate-income tax is apparently a good medium for generating government proceeds as long as the tax rate is not excessively high.

How does the depletion allowance fit into this scheme? Fig. 28 illustrates the effect of the depletion allowance. It is assumed that the depletion allowance that must be applied is 25% of the gross income, or 50% of the net income, whichever is the lesser.

In Fig. 28 an additional curve can be constructed to join the PV Depreciation/bbl curve—the new curve representing the Depletion Allowance/bbl is illustrated by the area between the PV Depreciation/bbl curve and the curves DP and PR. Curve PR must be chosen because in the upper part of the graph, 50% of the income is less than 25% of the gross income. In this part of the graph the effect of the corporate-income tax of 50% with depletion allowance on the economically-recoverable reserve is exactly equal to a corporate-income tax of 25%.

It illustrates that the depletion allowance is in fact simply a subsidy, administered through the corporate-income tax, for the benefit of petroleum producers. However, the depletion allowance is designed to allow a deduction for exploration costs that cannot be otherwise deducted for income tax purposes. These costs are usually small related to the total amount of depletion allowance.

Obviously, the corporate-income tax with depletion allowance is lacking as the most suitable method for earning the theoretical-maximum public revenue, because a large part of the rent can in fact be earned by the companies due to the depletion allowance. This is the reasoning behind the depletion allowance: allowing the companies a significant portion of the rent to stimulate future exploration.

Another medium could be the *progressive corporate-income tax*. Such a tax is highly selective. The scale must be based on the amount of taxable income issuing

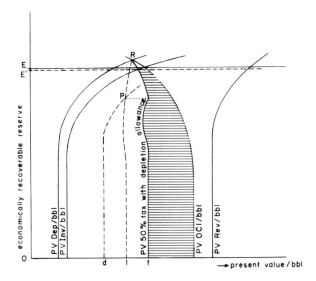

Fig. 28. Influence of the corporate-income tax with depletion allowance on the economically-recoverable reserve

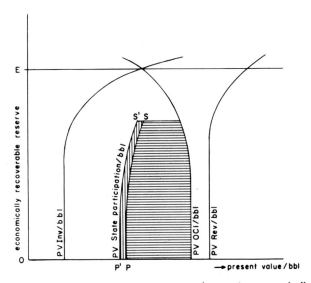

Fig. 29. Influence of state participation on the economically-recoverable reserve

from a concession in a year. A comparable income can be generated from a large field with a low profitability or a small field with a high profitability. Therefore, such a tax must be introduced with caution.

(5) *State participation*

The last element that will be analyzed is state participation. We will assume that the government participates after the discovery and that the state's debt to the company for its participation is paid with an interest rate lower than the internal-rate-of-return used in the graphs. Further, the supposition is made that the government does not participate in marginal projects. Consequently the graph—as is illustrated in Fig. 29—demonstrates no modifications in the upper part. In the lower part the result of a state participation of 50% is shown.

Firstly, this participation cuts into the surplus profits generated by the projects. The amount of profits earned by the government is shown horizontally hatched. Secondly, the companies' rents diminish because the state compensates incompletely for its participation by using an interest rate that is lower than that employed by the companies for the evaluation of the project.

Since the government does not participate in marginal projects, state participation never leads to a diminishing economically-recoverable reserve. This intimates that state participation is a suitable medium for procuring a large portion of the theoretical-maximum public revenue.

The conclusion which follows from this analysis is that state participation, the normal corporate-income tax and the sliding-scale royalties are suitable components of a petroleum law for obtaining a substantial portion of the theoretical-maximum public revenue for the state. The fixed royalties and the corporate-income tax with a depletion allowance are less suitable. The bonus and the surface duties are poorly adapted to achieving this goal. It must, however, be remembered that these preliminary conclusions are based on a theoretical model that excludes the geological risk.

The influence of several components of a petroleum law

Instead of using a single element, several components can be substituted for the previously discussed single ones to attain the maximum public revenue. Fig. 30 illustrates such a situation. The typical Tudor-arch-shaped arrangement of the PV Investment/bbl curve and the PV Operational-Cash Income/bbl curve is shown. The entire surface between these two curves represents the theoretical-maximum public revenue. It is assumed that this revenue equals 100 units, while the theoretical economically-recoverable reserve is 100%, because this is the largest reserve that can be recovered economically.

What happens when a particular bonus is applied by the government? Such a bonus is illustrated in Fig. 30 by line *a*. This line gives the PV Investment/bbl plus the PV Bonus/bbl for each barrel of production. The consequence of the bonus is that the economically-recoverable reserve diminishes to 97% (due to the new intersection *A*). This leaves a rent of 99.75 units to be divided between government and companies. The government is in this particular case earning 9.5 units, while the companies earn the remaining 90.25 units.

Now in addition to the bonus a fixed-surface duty per year is required for each project. This is illustrated in Fig. 30 by curve *b*. This curve represents the additions of the PV Investment/bbl, the PV Bonus/bbl and the PV Surface Duties/bbl. This new curve is intersecting the PV Operational-Cash Income/bbl curve at *B*. The new economically-recoverable reserve is now 93%. The divisible rent is now reduced to 98 units. The com-

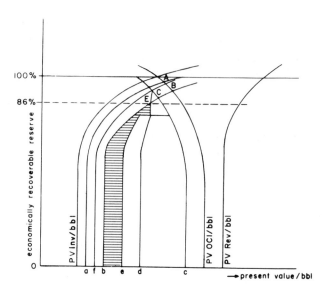

Fig. 30. Cumulative influence of the mining legislation on theeconomically-recoverable reserve

panies are earning 72.5 units and the government 25.5 units. This division of the rent is more favorable to the government than the previous one, but the application of surface rights produces minimized proceeds from the applied bonus. The government is now earning 8.5 units in bonus payments instead of the 9.5 units gained from a solitary bonus, and 17.0 units from the surface duties.

A royalty of 10% in addition to the previous requirements results in a new economically-recoverable reserve of only 89% (C). The divisible rent is now 96 units. The royalties to be earned account for 15 units, while the bonus earnings are reduced to 8 units and the income for the state from surface duties to 16 units. Total government proceeds are 39 units, leaving 57 units to the companies.

State participation may additionally result in curve d. Since marginal projects are excluded from state participation, the economically-recoverable reserve does not diminish and remains 89%. Consequently, the government proceeds from bonuses, surface duties and royalties remain unchanged. However, the proceeds from state participation are in this example substantial: 30.5 units, leaving for the companies a mere 26.5 units.

If in addition to these components a corporate-income tax of 40% of the taxable income is imposed, the economically-recoverable reserve diminishes to 86%. The divisible rent is 94 units. The earnings from the corporate-income tax are 14 units. The proceeds from the bonuses are now 7.5 units, from the surface duties 15.5 units, and from the royalties 14.5 units. The state participation remains unaffected at 30.5 units. This amounts to total government proceeds of 82 units, leaving 12 units for the companies. The obvious conclusions that can be drawn from this model are:

(1)   The addition of each new legislation instrument (with the exception of state participation) to the existing legal situation entails a decrease in the economically-recoverable reserve.

(2) The application of a new provision produces decreasing revenues from other elements in the mining bill. This follows from two consequences: the decrease in the economically-recoverable reserve and the direct effects of the elements on one another. An example is the decreasing amount of taxable income resulting from the enlargement of the requirements for bonuses, surface duties, or royalties.

Further, the decreasing economically-recoverable reserve results in a decreasing divisible rent. The decrease of the divisible rent, however, is not large if the economically-recoverable reserve equals a large part of the theoretical one.

An idea of how the different elements influence one another can be obtained from Fig. 31. In this figure a corporate-income tax is applied for the same fields as those analyzed in Fig. 30. The tax rate, however, is now 80%. This results in an economically-recoverable reserve which is 87% of the theoretical one, and a divisible rent of 95 units. The proceeds accruing to the government are 83 units. A tax rate of 80% has in this particular instance roughly the same effect as a composite of bonuses, surface duties, royalties, state participation and a tax rate of 40%. The reasons for the inclusion in petroleum laws of supplemental provisions in addition to the corporate-income tax become evident when the geological risk is introduced into the model in the following pages.

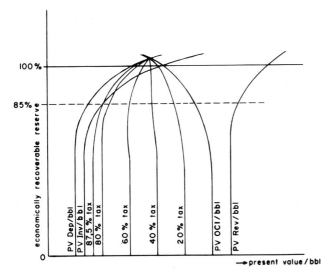

Fig. 31. Influence of a varying corporate-income tax on the economically-recoverable reserve

The critical-conditions pressure

How are government proceeds affected if, rather than the number and combination of elements, the extent of the application of a single element is varied—for instance, if the tax rate is varied from 0% to 100%? This will be studied for the tax rate in the cases given in Fig. 30 and Fig. 31.

Fig. 32 gives the government proceeds in units as a function of the tax rate. Case 1 is exhibited in Fig. 30. If no tax were applied, the government proceeds would equal 69.5 units due to the bonuses, surface duties, royalties and state participation. It can be seen in the graph that for a tax rate of 40% the government proceeds are 82 units as previously illustrated. These are, however, not the maximum proceeds that can be realized if the tax rate is varied. This maximum is reached when the tax rate is 61% and equals 84 units. At a tax rate of 80% the government proceeds are suddenly reduced to zero, owing to the decrease of the economically-recoverable reserve.

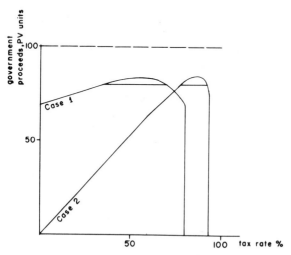

Fig. 32. Illustration of the principle of the critical range. Horizontal lines delineate areas in which more than 95% of the maximum public revenue is earned.

For case 2 the government proceeds are zero if the tax rate is zero. The maximum proceeds, about 85 units, are generated when the tax rate is about 87.5%. The government proceeds are once more reduced to zero when the tax rate slightly exceeds 93%.

It can be seen from the two graphs in Fig. 32 that when the amount of one element of a petroleum law is changed, with a consequent change in the conditions pressure for all the projects, there exists a point where the *maximum proceeds* for the government are reached. The amount of the maximum proceeds depends on the structure of the petroleum law. The point where the maximum proceeds are reached will be called the *critical conditions pressure*.

Another aspect emerges from the two graphs. There is a wide range surrounding the critical-conditions pressure, where the government proceeds are nearly impervious to a slightly higher or a slightly lower rate for the element (e.g., tax rate). The range where government proceeds exceed a certain percentage of the maximum proceeds will be called the *critical range*. In Fig. 32 the two critical ranges for case 1 and 2 are given, where government proceeds are no less than 95% of the maximum proceeds.

Due to the rapid decline of the economically-recoverable reserves, the government proceeds decrease rapidly as soon as the conditions pressure significantly exceeds the critical-conditions pressure. Such a sequence can be advantageously remembered when constructing petroleum law, since it demonstrates vividly how government proceeds can be lost when the conditions pressure becomes too high.

What determines the amplitude of the critical range? The wider the critical range, the less chance that government will loose proceeds due to a false establishment of the conditions pressure. First, the more elements a mining law incorporates, the more the risk is dispersed through the different elements and the wider the critical range—as can be seen by comparing cases 1 and 2 in Fig. 32. Second, the width of the critical range is determined by the shape of the PV Investment/bbl curve and the PV Operational-Cash Income curve. The more the Tudor-arch arrangement resembles a triangle, the broader the critical range; while this range dwindles the more convex the curves are toward the y-axis.

This model invites an inquiry into what conditions must be present for the government to earn the entire rent. This question is too theoretical to be of any value for further discussion. The simplest solution is a state participation of 100%, meaning nationalization of the complete petroleum exploration and production process. In this case, however, the state must in addition bear all the risks alone.

Additional remarks about the model

Before any definite conclusions can be drawn from this model a number of additional remarks must be noted to make a start at adapting theory to reality. Four of the assumptions made thus far must be modified:
(1)   The assumption about the Tudor-arch-shaped arrangement of the curves.
(2)   The assumption that only a homogeneous fluid, petroleum, was being dealt with.
(3)   The assumption that the price was the same for each barrel of output.
(4)   The implicit assumption that the state would interpret the divisible rent on the basis of the same interest rate as the company.

(1)   *Exceptional cases*
The Tudor-arch arrangement of the PV Investment/bbl curve and the PV Operational-Cash Income/bbl curve was based on the supposition that the larger fields would be on the average the more profitable ones. This is by no means certain. Large fields with poor productivity per well may occur; or fields may lie in deep water far from the shore, or in geographically remote areas. In such eventualities, rent on the fields will be minimal, or production may even become marginal. Such a situation is outlined in Fig. 33.
When large but marginal fields lie in an area of jurisdiction, the shape of the PV Investment/bbl curve and the PV Operational-Cash Income/bbl curve will be that of an Ogee-shaped arch. If this situation occurs, almost any type of mining legislation—such as a simple 50% corporate-income tax—will result in the loss of considerable reserves. In Fig. 33, such an income tax results in a loss of up to 72% of the theoretical economically-recoverable reserves. If the government wants these reserves to be produced, for instance, to attract foreign exchange, the mining legislation should be enriched with potential exemption from payments to the government in exceptional cases. Such a possibility costs the government little, because

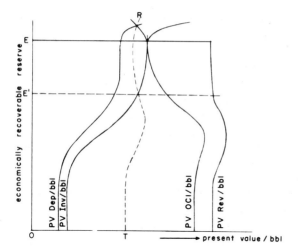

Fig. 33. Illustration of an exceptional case for the distribution of project profitabilities in a given area

the rents on large but marginal reserves are minimal, while the benefits to the country from their production may be abundant.

(2)   *Oil and gas*

Until now a homogeneous fluid, petroleum, was assumed to fill the different reservoirs. In practice, however, there will be a number of different type fields: oil fields, gas fields, oil fields with a gas cap, gas fields rich in natural-gas liquids, etc. When oil and gas are produced from the same field, the "joint-cost" problem arises. The distribution of costs over gas and oil is a topic that is treated in different ways. SMITH and BROCK (1959) mention four methods:

(*a*)   According to certain physical units. The oil and gas can be compared on an energy basis. Costs can be split over comparable energy quantities. Other physical units such as the weight of the respective fluids can also be used.

(*b*)   According to relative costs. It is possible to study separate oil and gas reservoirs and draw conclusions for the separate average costs of producing gas and produce oil.

(*c*)   According to "sales realization". The costs are allocated based on the revenues earned from production.

(*d*)   According to combinations of these methods.

A new method has been introduced by TAHER (1966). He proposes an allocation of costs according to the chemical composition. No allocation problems, at least for the direct costs, arise when single oil or gas fields are discovered.

The revenue at the terminal for oil and for gas depends heavily on the geographical distance from the main consumption centers. For oil there exists a world market, as illustrated in Chapter III. Oil near consumption centers in the U.S.A. and western Europe can be sold for substantially high prices, as U.S.$ 3.30 per barrel in the

U.S.A. Oil far from these centers is sold for considerably less; for instance the "posted prices" in the Middle East are U.S.$ 1.80 per barrel.

Within the limited area of jurisdiction in a small country, or for offshore concessions confined to a single area, the price differences between different oil fields are slight. A certain variation in price may originate from the quality of the oil.

There is as yet no world market for natural gas (Chapter III). In western Europe and the U.S.A. natural gas can be profitably sold. Natural gas in the Middle East, however, is almost worthless. The discovery of a gas field in these areas can virtually be treated as a dry hole unless the gas can be used for recovery of condensate. The differences in price within a rather small area may be significant due to the high transport costs.

If an area consisted entirely either of oil fields or of gas fields, the familiar Tudor-arch-shaped configuration of the PV Investment/bbl and the PV Operational-Cash Income/bbl would be the most likely arrangement. Relatively low PV Average Costs/bbl can be expected for the large fields, while relatively high PV Average Costs/bbl will occur for the smaller fields.

If an area, however, consists of a mixture of oil and gas fields, the curves will in all likelihood be highly irregular—if, on whatever basis (energy content, volume, etc.), the gas and oil fields are grouped on the same graph. A petroleum law applicable to all fields is therefore pointless, suggesting the development of a separate analysis for oil and gas. Oil fields containing considerable amounts of gas or a separate gas cap can be treated in such a way that natural gas is converted for the purpose of calculation into equivalent barrels of oil on the basis of the same energy content, or with any other conversion method. Gas fields whose primary product is natural gas, but which contain considerable amounts of natural-gas liquids or oil, can be similarly treated by translating non-gas holdings into natural-gas terms.

Curves showing a greater variety of costs and revenues for oil as well as for gas result from this analysis. Separate conditions can be formulated for oil and for gas, eventually making provisions for situations such as gas caps or natural-gas liquids. Oil and gas legislation may differ considerably following governmental appraisal of this analysis, especially in producing countries far from consumption centers.

(3) *Prices*

Prices were regarded as being constant in the model. The distinction between oil and gas has already introduced a price differential into the model. A number of other aspects of prices are investigated in this section, including the effects of price changes, posted prices, conservation measures and monopolistic attitudes.

Price changes. The effect of price changes is exactly comparable to the effect of a fixed royalty. For instance, a price decrease from U.S.$ 3.30 per barrel to U.S.$ 2.97 per barrel has precisely the same consequence as a royalty of 10%. However, the "royalty" is in this case earned by the consumer, instead of the state. An eventual lower price for oil in the U.S.A. following liberalization of import policy would therefore result in the elimination of marginal fields—in this case, only those on the U.S.A. east coast.

Posted prices. Fig. 34 illustrates the effect of a tax calculation employing posted prices, while the oil must actually be sold for lower realized prices. If the theoretical economically-recoverable reserve was OE on the basis of posted prices, the decline in the reserve can be constructed by moving the PV Operational-Cash Income/bbl curve, together with the tax curve RT, a certain distance to the left. This distance should equal the difference between the present value of the posted price and the present value of the realized price. In this case the economically-recoverable reserve has declined to OE' due to the price cut, but the new tax curve R'T' results in a further cut of the economically-recoverable reserve to OE''. This example demonstrates that the use of posted prices or tax-reference prices in taxable income

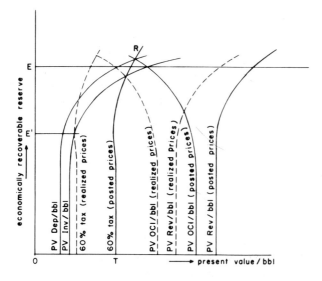

Fig. 34. Influence of posted prices on the economically-recoverable reserve

calculations may entail a considerable loss of reserves, when a wide variety of PV Average Costs/bbl prevails among the different fields. This loss is comparable to that emanating from the use of very high tax rates. The danger exists, however, only if there is a considerable difference between the realized price and the posted price. If this difference is small, the use of posted prices—higher than realized prices—is a useful medium for enlarging government proceeds.

Conservation measures. Intensive schemes of market control, in the U.S.A. in particular, seek to regulate the oil supply. The yearly output of profitable fields in Texas and Lousiana is considerably restricted, while the marginal wells may produce at full capacity. This is an artificial restriction of supply and it leads to the establishment of high price levels. The results of arbitrary conservation measures on our model can be seen in Fig. 35. The price of the marginal wells is increased by the conservation measures. The present value of the price, however, for the large and prolific fields will be lower, due to the slow production rate; while the costs per

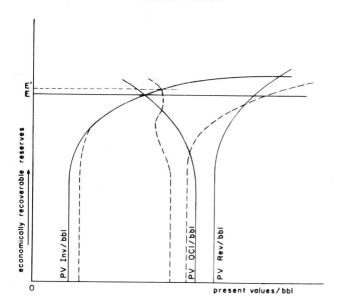

Fig. 35. Influence of conservation measures on the economically-recoverable reserve

unit production are higher. The conservation measures give rise to a slightly increased reserve OE', but a decreased divisible rent. It is therefore questionable to what extent such conservation measures profit the state. The average expected governmental proceeds diminish substantially with such conservation measures, while consumers are paying high prices. The only benefit is a slightly increased economically-recoverable reserve and a larger capacity for potential oil production (if installations are maintained to produce at a higher rate than the allowable one).

Monopolistic attitudes. Another method of limiting the market can be by restricting output to a few large fields, while enjoining the marginal ones from any production. Especially in gas markets, this monopolistic holding brings about high prices. These higher prices provide remunerative direct proceeds for the state, but they penalize the consumer. WOLFF (1964), or Chapter III of this book, will afford a more elaborate study of this question.

(4) *The interest rate used by the state*
The previous discussion allows us to draw a number of conclusions concerning the provisions of mining legislation that must be in effect to approach the maximum public revenue. These proceeds, however, are discussed in terms of an interest rate that equals an acceptable rate-of-return for the companies. It is not necessary that the government interpret the returns from the mining operations with the same rate-of-return as do the companies. The question might arise if the application of another interest rate affects the critical-conditions pressure and consequently the conclusions that can be drawn from the model. Since the shapes of the curves are

partially determined by the interest rate, a change in this rate alters the critical-conditions pressure. In the case of the normal Tudor-arch arrangement of the curves it is debatable whether the change is very large. In Fig. 36 this Tudor-arch grouping is shown. It is assumed that the PV Investment/bbl curve is unaffected by the government-interest rate. The government rate is supposed to be considerably lower than the company rate. The PV Depreciation/bbl curve and the PV Operational-Cash Income/bbl curve, however, change positions. Suppose the change is that indicated by the solid lines. The PV value of the tax earned at the original critical pressure of 87% is greatly increased by the lower interest rate. Such proceeds, however, are the maximum possible in this new situation. This follows on the one hand from the rapid decline of the economically-recoverable reserve, after the

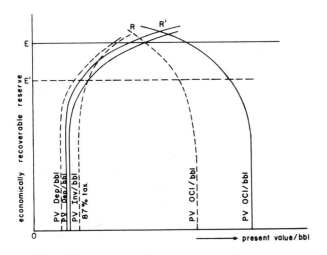

Fig. 36. Influence of the interest rate applied by the government, on the total amount of government revenues

maximum proceeds have been reached. On the other hand, few additional proceeds can be reckoned on from a lower tax rate, because the rent of the additional reserves is low in comparison to the more profitable reserves in the graph. In the case of Fig. 36, the critical-conditions pressure is probably at a tax rate of 2 or 3% lower than 87%. With the same tax rate at this government interest rate, the proceeds remain well within the critical range. Note that the economically-recoverable reserve does not change with the manipulations of the government interest rate. The companies decide whether or not a project is attractive. Thus, the economically-recoverable reserve remains the same whatever interest rate the government applies for its scrutiny of the maximum public revenue. The reserve only changes when the government alters the conditions pressure. Also under study is whether the same critical-conditions pressure can be composed in a way more profitable to the state. At a high interest rate the value to the government of the proceeds from state participation and corporate-income tax is much lower than at a low interest rate. The value of the initial bonus remains unchanged, and the value of the surface rights and royalties falls in an intermediate position. The way the government obtains its interest rate for the calculations is beyond the scope of this study. It is closely related to the country's policy and to development planning in the case of developing countries.

The net-present-value model: conclusions

The conditions that a petroleum law must meet to maximize the public revenue in a given area are summarized below. Compilations are based on a fixed legislative framework (which remains constant throughout the lifetime of the project) and the absence of geological and other risks.

(1) Since the costs and the revenues from oil fields differ widely from those of gas fields, it is useful to split mining law into separate provisions for "gas" and for "oil".

(2) Each petroleum law must provide tools for handling exceptional cases, such as production from large, nearly marginal fields, since the model incorporates the general assumption that large fields are normally profitable.

(3) The maximum proceeds from a particular mining law are obtained at the critical-conditions pressure. If another conditions pressure is applied, given components can better be fixed at amounts which are too low than too high, because the economically-recoverable reserve decreases rapidly when one or more elements are fixed too high, with a consequent decline in government proceeds.

(4) If a large variety of oil or gas fields is anticipated, the critical range for establishing a certain element within a given mining law is wide. This range is narrow when a single field or only a few similar fields are producing.

(5) If more than one element is employed—as is usually the case—it must be remembered that the application of each new element reduces the economically-recoverable reserve and accordingly the proceeds from other components of the mining legislation.

The only component of a mining legislation that does not necessarily diminish the economically-recoverable reserve is state participation.

(6) State participation, sliding-scale royalties, and the corporate-income tax are well suited to generating the maximum public revenue. The fixed royalties and the corporate-income tax with a depletion allowance are less appropriate, while the surface duties and bonuses are totally unfitted to achieving this goal.

The influence of geological risk

To discover oil and gas a company must be willing to invest in an exploration program. Mining legislation must encourage the company to take the risk involved in investment in an exploration program. If the company is not emboldened to explore, the profitable production project is not discovered and consequently not produced. The greater the exploration activity the greater the chance that profitable projects will be discovered. The decision to invest in an exploration program can be made with the help of a calculation of the expected-monetary value, as is described in Chapter IV. This method of investment analysis will be used for the discussion in the following section.

The term "exploration activity" is vague. It is difficult to specify whether exploration activity is extensive or limited within a particular area. The number of seismic crews, or drilling projects, or number of leased blocs is, of course, an absolute measure of activity, but since both government and companies have insufficient geological knowledge, no one can determine whether exploration activity is relatively too large or too small.

Expected-monetary value and conditions pressure

In Chapter V the expected-monetary-value concept was used to illustrate the influence of mining legislation on the profitability of a single project. In this chapter the concept will again be used for an analysis of the government's standpoint.

The formula for the calculation of the expected-monetary-value, used in Chapter V, can be here given in a simplified form as:

> $EMV =$ (probability of success) x (expected part of the project rent) $-$ (probability
> of a dry hole) x (expected loss on pre-discovery investments)

or:

> $EMV = p$ x (expected part of project rent) $- (1 - p)$ x (dry hole costs)

The only reasonable certain factor in this equation is the expected loss on pre-discovery investments. Although exploration costs may vary among the different projects, they can be estimated rather closely. Each project is attractive as soon as the expected

part of the project rent $= \dfrac{(1 - p)}{p}$ x (dry hole costs)

If the interest rate for the calculation is not fixed at a level where dry-hole risk is included, the dry holes must be financed from the rent on the project, which is the quasi rent. The right hand side of the equation gives the minimum quasi rent that is acceptable to the company.

Fig. 37 illustrates the foregoing formula. The x-axis gives the probability-of-success. The y-axis gives the minimum-expected-project rent that is acceptable to the company for each probability-of-success. This "expected-project rent" is expressed in the number of dry holes, or more exactly, in the number of expected pre-discovery losses.

The lowermost curve in Fig. 37 illustrates a situation in which no payments are required by the government. For instance, if the probability-of-success is 0.2, the expected-project rent must equal at least the value of 4 dry holes.

When payments to the government are required, a certain portion of the expected-project rent must be paid to the government. This part was previously defined as the conditions pressure. In this case the conditions pressure must obviously be defined in terms of the expected-monetary-value. Different curves are given in Fig. 37 for various conditions pressures. For instance, the expected-project rent must equal 8 dry holes if the

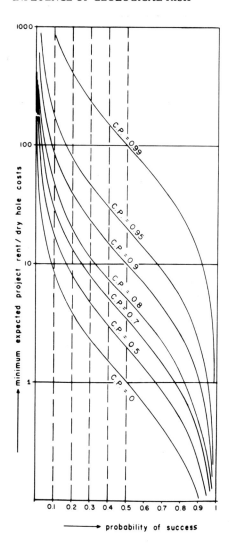

Fig. 37. Probability-of-success and required expected project rent

probability-of-success is 0.2 and the conditions pressure is 0.5.

As soon as the conditions pressure becomes somewhat high, for instance 0.9, the expected-project rent must be excessively high to make an exploration project profitable. Assume, for instance, dry-hole costs in the North Sea of U.S.$ 3.– x 10^6. If the probability-of-success is 0.1, the expected rent (present value) on the project must be U.S.$ 270.– x 10^6. This means that the company must assume at least a fair chance of finding a Leman-type gas field or equivalent oil field, because the given figures are *average* expectations.

A typical attribute of the expected-monetary-value technique is its leveling aspect. It is most unlikely that the actual net-present value of a project will turn out to be exactly the expected-project rent. The expectation mainly encompasses small chances of very large discoveries and greater likelihood of moderate or marginal discoveries. If a very profitable field is discovered, the expected-project rent would be considerably less, since it also included possibilities for moderately profitable projects. On the other hand, the expected-monetary value of each realized drilling project must be assumed to have been positive when the drilling decision was made; consequently, the expected-project rent must have included the value of several dry holes. If a dry hole is drilled, an expected-project rent higher than the actual net-present value was originally calculated. This means that if two distributions, the distribution of the actual net-present values of the projects and the distribution of the expected-project rents, are compared, the range of the latter distribution will be considerably narrower than the range of the former.

The range of the expected-project rents, expressed in number of dry holes, for an offshore area falls somewhere between 5 and 500 holes. It is not likely that a project will be accepted, whatever the probability-of-success, if it is not possible to drill another 5 dry holes, if it proves to be successful. Apart from the economic, technical and geological risks, some special hazards—such as the occurence of a disaster like the oil slicks from the Santa Barbara Channel or Louisiana offshore—are not yet included in the calculations. It is worthwhile to avoid even such small chances of great disasters by eschewing marginal projects.

On the other hand, average expectations of projects with a rent of 500 times the offshore dry-hole costs are geologically improbable.

The range of the expectation for most offshore projects would probably fall between 10 and 100 times the dry-hole costs. This range will roughly hold for most offshore areas of the world. On land, the upper limit of the range will be considerably higher because of the lower drilling and development costs. The lower limit will be lower, since risks additional to geological ones are less prevalent.

Geological risk and conditions pressure

If it is assumed that the expected-project rent ranges between values equal to 10 and 100 holes, the two other variables, the geological risk and the conditions pressure, can be studied. The conditions pressure clearly must be lower the lower the expected probability-of-success in realizing "exploration activities" for a given area. This relation is studied in Fig. 38.

Three different cases are given:
(1)   A high probability-of-success (p ranges from 0.25 to 0.5).
(2)   A moderate probability-of-success (p ranges from 0.1 to 0.25).
(3)   A low probability-of-success (p ranges from 0.01 to 0.1).

For these three situations, what will be the conditions pressure that the government can apply to all projects, and still induce the companies to explore sufficiently?

(1)   Fig. 38a shows the situation for a generally expected high probability-of-success. When the conditions pressure is lower than 0.7, all the projects will be acceptable.

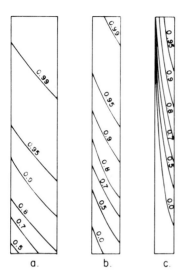

Fig. 38. Allowable conditions pressures for three ranges of probability-of-success, a — high probabilities-of-success; b — moderate probabilities-of-success; c — low probabilities-of-success. This figure
gives details of Fig. 37. Range: 5-500 times dry hole costs.

When the conditions pressure rises beyond 0.99, all the projects will be unacceptable to the companies. A conditions pressure of 0.8 will not seriously affect the
exploration activity, but at a conditions pressure of 0.9, a number of projects will
become unattractive. A sound government policy, therefore, would be to keep the
conditions pressure between 0.8 and 0.9 for a given situation.

(2)   Fig. 38b gives the conditions for a generally expected moderate probability-of-
success. As long as the conditions pressure is less than 0.1, all the projects in the
range are attractive; while above 0.97 no profitable projects exist. However, as long
as the conditions pressure remains below 0.5, only a few projects will be dropped
by the companies. Many exploratory ventures become unattractive above a conditions pressure of 0.7. Consequently, government policy should aim at a conditions
pressure between 0.5 and 0.7.

(3)   Fig. 38c illustrates the case of a generally expected low probability-of-success. Even
if the conditions pressure is 0, a large number of projects will be unattractive. All
the projects are unprofitable if the conditions pressure exceeds 0.91. If the government wishes to induce the companies to explore actively, the conditions pressure
should be below 0.5.

The sequence of exploratory trials

Geological expectation is continuously changing. For instance, as soon as the petroleum law is put into effect and the number of companies demonstrating interest in the
entire area or certain blocs becomes known, the geological expectations for the area may
change. If a very active interest exists for the area, more geologists will be inclined to

estimate a higher probability-of-success; because no single company can unravel all the geological mysteries of a certain area, promising outcomes of the investigations of other companies may enlarge the faith in their own results.

As the results of the first exploration wells become available, however, the picture changes radically. The first exploration returns may be discouraging. In this case projects based on a similar geological model, but not yet tested, become similarly risky. However, if the first results are promising, the same untested projects may become of a moderate or low risk. This general rule can be applied only on projects contained within a certain geological model. As soon as different geological models are introduced as the bases for exploratory activity (generally the case), the effect of the results within one geological model on the overall exploration activity is rather complicated, as is explained in Chapter IV.

The possibilities for profitable projects beneath a certain geographical area are not un-limited. In the first phase of offshore exploration, exploration wells are drilled according to seismic evidence of large structural traps. The number of such traps in a certain area is, of course, limited. When the larger traps are tested and they contain no profitable oil or gas, the enthusiasm for testing the smaller traps will gradually cease.

This does not mean that the possibility of a new phase in exploration activity is eliminated. New seismic techniques make possible the detection of large structural traps formerly too difficult to place. The discovery of the Slochteren gas field was partially a result of such improved seismic techniques on land. Further, when a number of explora-tion wells has been closely studied, evidence may appear for other types of traps, as stratigraphical ones that may contain large oil or gas fields.

In the first phase of exploration offshore, activity gradually diminishes following negative results of tests on the large structural traps, and when different geological models have been tested. The risk of a diminishing exploration activity is particularly large if the probability-of-success is low. The chance that a number of dry holes will be drilled at the immediate beginning of an exploratory trials sequence is in this case statistically high.

To give an idea of the extent of this possibility, a set of Bernouilli trials can be analyzed. It turns out that the chance of ten successive dry holes if $p = 0.1$ is 0.35; and if $p = 0.25$ this chance is 0.05 or 5%. The chance of ten successive dry holes is large enough to be realized in such a situation even when it is impossible to speak of a mathematically pure Bernouilli condition. On the other hand, ten successive dry holes may be totally destructive of further exploration activity, especially offshore. Consequently, there is a purely statistical chance that a large number of successive dry holes will occur, but that the original estimation of the geological risk was afterall justified, since continued explo-ration would have revealed profitable oil or gas fields.

If this situation forces a halt to exploratory activity, we can say that the company has fallen into the *dry-hole trap*. The larger the initial exploration activity the smaller the chance that the dry-hole trap will prevail.

The composition of the conditions pressure

As was explained previously, the same conditions pressure for a project can be actualized by different sets of petroleum laws. A corporate-income tax of 70% will, for instance, exert a conditions pressure that will be certainly higher than 0.7 (due to the difference in taxable income as calculated by the government and the project rent as calculated by the company). A government participation of 70% will also result in a higher conditions pressure (due to the expected differences in interest to be applied for the compensation to the companies). For profitable fields the differences between the tax rate, state-participation rate, and conditions pressure will be slight; for the less profitable fields these differences may be greater, as is diagramatized extensively by the net-present-value model. For other components of mining legislation similar conclusions will hold; but without an exact calculation for a specific case, no figures can be given. For instance, a bonus equalling the costs of 5 dry holes, will yield a conditions pressure of approximately 0.5 if a project rent equal to 10 dry holes is anticipated; while the same bonus will result in a conditions pressure of only 0.05 if 100 dry holes could be paid from the rent.

The general conclusion drawn from the net-present-value model was that a petroleum law must be *highly selective*. This conclusion is equally applicable to the expected-monetary-value model as will be demonstrated in Fig. 38. If the probability-of-success is expected to be 0.1 (Fig. 38b), a conditions pressure of 0.9 is permitted for projects with an expected-project rent equalling more than 90 dry holes. If the project rent is expected to equal only 18 dry holes, the conditions pressure should not be more than 0.5. Consequently, if the same mining legislation can require a conditions pressure of 0.9 for the very profitable projects, and of only 0.5 for the less profitable ones, both situations will be deemed favorable by the companies. The exploration activity will therefore be high with selective mining legislation.

Selectivity in mining legislation is only possible with regard to the ultimate-project rent, and can exert little control upon the expected probability-of-success. This implies that even with a highly selective petroleum law, a range of possible conditions pressures will occur due to the expected range of the probability-of-success. For instance, if the expected-project rent equals 18 dry holes and the probability-of-success ranges from 0.1 to 0.25, the conditions pressure could theoretically range from 0.5 (for $p = 0.1$) to 0.83 (for $p = 0.25$).

Due to the unique risks offshore, a company is unlikely to accept a project that becomes marginal following payments to the government. If the special risks are assessed to equal 5 times the dry-hole costs—as previously arbitrarily assumed—the actual evaluation of the project would take place on the basis of an expected-project rent not of 18, but of 13 times the dry-hole costs. Hence the conditions pressure must be lessened to induce the companies to explore.

Of course, there is no need to fix the petroleum law in such a way that all the projects within the quadrangles of Fig. 38 are profitable. A slight withdrawal will not affect the overall-exploratory activity; but when slective-mining legislation is fixed in such a way that numerous projects from the upper to the lower end of the quadrangle become unacceptable to the companies, the government has pressed too far.

This discussion has given an overview of how the conditions pressure can best be applied in the selective-petroleum law. Now the six conclusions drawn from the net-present-value model require closer study.

(1)   The distinction between oil and gas fields is equally valid for the expected-mone-
      tary-value model. When a general insight has been generated into the differences in
      revenues and costs, the generally expected-project rents can be studied for oil and
      for gas. The mining law can establish distinct conditions pressures for "oil" and
      "gas". This is an important step toward a greater selectivity in mining legislation.

(2)   If a company assumes a slight possibility of the occurrence of an exceptional
      condition, it may considerably affect its propensity to explore if no special condi-
      tions are included in the petroleum law to meet this situation.

(3)   The order of magnitude of the critical-conditions pressure can be established in the
      expected-monetary-value model with studies based on the relation between
      expected-project rent and probability-of-success, as previously illustrated. A more
      detailed analysis follows. The conclusion that the conditions pressure can be better
      established too low than too high is equally applicable for this section, because an
      unwarrantedly high conditions pressure will considerably reduce the propensity to
      explore. A critical-conditions pressure as such cannot be established for the
      expected-monetary-value model because the results of the exploratory efforts are
      entirely uncertain.

(4)   If a large variety in oil and gas fields is expected, the variety in the expected-project
      rents will generally be abundant. This implies that the conditions pressure can be
      established within wide limits.

(5)   The number of acceptable projects will decrease as soon as a new element in the
      petroleum law is added to existing ones, leading to a decrease in exploration activi-
      ty. An interesting distinction can be made between the net-present value and the
      expected-monetary-value model—state participation in the latter case may lead to-
      ward a decrease in the number of acceptable projects. If no discovery is made, the
      risk is usually borne entirely by company, but if the discovery results in a consider-
      able positive cash flow, state participation is applied. This effect upon the desirabili-
      ty of a project has been previously illustrated in Chapter V.

(6)   The conclusion that the maximum public revenue must be reached with the state
      participation, sliding-scale royalties and the corporate-income tax is consistent with
      the conclusion in Chapter V that these three elements are all post-discovery provi-
      sions exerting relatively less influence on the expected-monetary value than pre-
      discovery elements like the initial bonus. Furthermore, these elements are highly
      selective and therefore do not reduce exploration activity excessively.

The conditions pressure to be applied

These preliminary inquiries invite a more detailed investigation of the conditions
pressure. The situation hypothesized in Fig. 38 will be analyzed more completely below.

When mainly high probabilities-of-success ($p = 0.25$ to $p = 0.5$) are posited, the
conditions pressure should accordingly be high. As previously indicated, the conditions
pressure can fall somewhere between 0.8 and 0.9 without seriously affecting the number
of projects acceptable to the companies. Thus, for instance, a corporate-income tax of
50% together with a state participation of 50% will engender virtually no withdrawals
from the interesting projects. It leaves, however, a number of projects that could have

borne a heavier conditions pressure. To fill this gap, a sliding-scale royalty can be intro-duced, amounting to 20% for the very large fields. Low initial bonuses and surface duties can be added to the list.

With moderate probabilities-of-success ($p = 0.1$ to $p = 0.25$) the conditions pressure must be more selective. The implication is that for the less attractive projects only a corporate-income tax of 50% will suffice. For the better projects state participation can be maintained, but it is good policy to stress that state participation will occur only when demonstrably profitable projects have been found. A sliding-royalty scale is equally adaptable, but cumulative royalties for the less attractive projects should be low, for instance not exceeding 10%. Low bonuses and surface duties can be added.

When probability-of-success is low ($p = 0.01$ to $p = 0.1$), the mining legislation can be better fixed at a corporate-income tax of 50%, with state participation excluded from the law since the chances of finding projects able to bear a conditions pressure of 0.8 will be minimal. The sliding-royalty scale should be moderate and should probably not exceed 10% for all projects. Since geological risks are high it is advisable to introduce a liberal-taxation policy. For instance, a depletion allowance can be inaugurated to reduce the effect of the 50% tax rate; or special tax-write-offs can be granted. Such a policy would increase the allure of exploration projects and would avoid the dry-hole trap.

The proposed legislation for the various geological conditions is of course only applicable to the model given in Fig. 38. This will be, however, an acceptable model to apply for oil and gas in offshore areas adjacent to industrialized regions such as the North Sea, the Adriatic Sea, or the Gulf of Mexico.

The level of costs and revenues will be highly dependent upon the area in which the exploratory ventures are situated. For instance, in Alaska, or offshore Africa—far from industrialized regions—the differential will be less profitable to the companies. Costs are higher on the average, due to higher outlays for equipment, repairs and highly qualified technical labor; while revenues can be expected to be lower due to the absence of nearby markets. These less favorable economic conditions will result in a narrower range of expected-project rents. Firstly, the upper limit is moving downwards in the graph, due to the lower revenues and higher costs. Secondly, the lower limit can be expected to move upward when in addition to the normal extraordinary risks, such as pollution or other accidents, political risks are included. Described in more literary language, the oil com-panies are hunting only elephants in offshore Africa, while in offshore Europe, deer are also attractive.

In industrial areas onshore, oil companies are even hunting rabbits. Onshore the relation between costs and revenues is more favorable than offshore and the range of the expected-project rents will be wider. The onshore Africa range is probably comparable to offshore conditions in Europe.

Apart from the differences in the width of the ranges, the differences between the profitability of oil and of gas projects increase for less industrialized regions as previously outlined.

Geological risk and the government attitude

In introducing new mining legislation into a previously unregulated area, government must make the weighty decision of the extent to which it wishes to involve itself in the geological risk. If the government ascribes significant weight to post-discovery elements in its mining legislation, it risks receiving almost no income if the exploration proves to be entirely unsuccessful. If, on the contrary, government requires bonuses, it is ensuring secure proceeds that can not be affected by the exploration results. The element of risk to the government is a strong argument for the application of bonuses, and to a lesser extent surface duties. However, the application of a fixed bonus and of fixed- or rising-surface duties adversely influences the propensity to explore.

There exists, however, a means of converting the bonus as well into a selective piece of mining legislation—through bidding. Companies may be left the option of determining how much they want to pay for a certain bloc. This amount can be determined by auction or by sealed bidding. The result in both cases is to offer the bloc or concession to the highest bidder. The system of sealed bidding has been successfully applied in the U.S.A.; in 1968 alone the amount spent on offshore-lease bonuses was U.S.$ 1.4×10^9 (ANONYMOUS, 1970). These bonuses were paid above and beyond other outlays due to state and federal governments, such as royalties and corporate-income tax.

The system of sealed bidding can be successfully applied when:

(a)   the number of companies expected to bid on the blocs is relatively large;
(b)   the government is indifferent as to which company obtains the concession.

Both conditions prevail in the U.S.A. Considerable interest is exhibited by the whole corps of large petroleum companies for every offshore area, while the government is unconcerned as to who obtains the concession because the great majority are domestic firms. Control consequently remains in American hands, while no one company can presently be expected to reach a monopolistic position within the U.S.A.

For small or developing countries, however, the mentioned conditions seldom occur. The number of companies is usually small, especially when little interest by outside petroleum companies is displayed in the area. Further, the countries are too small simply to grant the concession to the highest bidder. It could result in the entire offshore area falling into the hands of a single company, making the national petroleum policy entirely dependent on the moves of this company.

The bidding system is hence an inappropriate legislative tool for most countries. In such situations it is normally better to grant concessions to a number of different companies, if possible, and to negotiate for the companies to accord additional benefits to the state. Such "benefits" might include the obligation on the part of the company to spend a large sum on exploratory activity and/or the obligation to invest in other sectors of the country's economy—such as the extension of refinery capacities, establishment of new industrial activities, or cooperation with the state in other mining or technical projects.

There exists another argument, apart from the geological risk, for the inclusion of bonuses and surface duties in the petroleum law—that is, to maintain a regular cash flow of the government proceeds. If the exploration is successful, the maximum public revenue will be earned when state participation, corporate-income tax and sliding-scale royalties are applied; but this implies that the government is deprived of all proceeds from the mining operations until production starts and royalties can be earned. This time lapse may be several years to more than a decade. In some instances it is hardly possible for governments, especially in developing countries, to wait so long for benefits.

Small proceeds in the form of bonuses, and annual-surface duties may in such cases cover temporary needs. The inclusion of bonuses and surface duties in petroleum law may, therefore, be beneficial to the state, due to the aspects of geological risk and the expected-cash flow for the government.

Another aspect of the governmental attitude towards geological risk remains to be discussed. From three different cases shown in Fig. 38 it can be seen that the establishment of a certain conditions pressure for all projects becomes more and more difficult the lower the probability-of-success. For the low-risk situation the conditions pressure should be between 0.4 and 0.99, for the moderate risk between 0.0 and 0.99, and for the high risk between 0 and 0.97. Especially in the latter two cases the government is confronted with an unpleasant choice: either it puts a petroleum law with a high conditions pressure into effect (risking very low or non-existent exploratory activity) or it establishes a low-conditions pressure which results in the companies' earning considerable portions of the rent when exploration proves successful.

The effects of this choice can be softened by releasing only a limited area for exploration under the new petroleum law. This area can be offered under terms barely sufficient to induce a substantial number of exploratory trials. If such an exploration succeeds, the probability-of-success for the remaining area will be placed considerably higher and firmer terms can be established. If exploration is unsuccessful, the government can offer the remaining area under more liberal terms.

Conditions for achieving the maximum public revenue

The analysis is now developed. The net-present-value model and the expected-monetary-value model have been discussed. From these investigations a number of conclusions can be drawn that are important for the future structure of petroleum law—if the government's target is solely the earning of the maximum public revenue.

A general conclusion emerging from all the discussions is that the petroleum law must be highly selective, to earn as large a part as possible of the true rent of each project and to still keep as many projects as possible attractive to the companies. This selectivity can be structured by:

(1) Applying different regulations in the petroleum law for oil and gas.
(2) Providing for financial arrangements to cover exceptional cases (occurrence of large but marginal fields).
(3) Giving the most weight to elements such as corporate-income tax, state participation and sliding-scale royalties.
(4) Regulating the conditions pressure according to the generally expected probability-of-success. The conditions pressure should never exceed 1, and must account for the distinction between offshore and onshore areas, while the general economic setting must be included as well. For offshore areas adjacent to industrialized countries, or onshore areas in non-industrialized countries, the following general tendency can be noted:
   (a) with a generally expected high probability-of-success $(0.25 < p \leqslant 0.5)$: 50% corporate-income tax, 50% state participation, and sliding-scale royalties up to 20% of the per-barrel value;

(*b*)   with a generally expected moderate probability-of-success ($0.1 < p \leqslant 0.25$): 50% corporate income tax, state participation only in rich discoveries, sliding-scale royalties;

(*c*)   with a generally expected low probability-of-success ($0.01 < p \leqslant 0.1$): 50% corporate-income tax with depletion allowance or special tax deductions, and a moderate sliding-scale royalty not exceeding 10% of per-barrel value.

With regard to the government's attitude toward geological risk two other conclusions can be made:

(5)   The application of low bonuses and surface duties is advisable to reduce the risk of a zero income and provide an early cash flow.

(6)   In the case of high geological risk the petroleum law can be restricted to a limited area.

All these conclusions are based on the fact that the companies accept a mining bill which is applicable to the projects during their entire lifetime. This is generally untrue. What follows from a change in mining legislation?

Changing the mining law

A change in mining legislation during the lifetime of a certain project can have two causes. Portions of the mining law may be dependent on other legislation that may change, for example, with general economic conditions. And other specific circumstances may change so radically that the companies or the governments express the wish to negotiate the terms anew.

Changes due to general economic conditions

Frequently, changes due to general economic conditions accompany an adjustment of corporate-income tax if this tax is not established by the petroleum law but in the general tax legislation of the country. A change in general economic conditions may result in an alteration of the corporate-income-tax rate. The change in tax rate is in this case usually related to the welfare-economic goal. If the country is confronted with considerable inflation, diminishing economic activity may be sought; a higher tax rate may meet this need. A higher tax rate will engender reduced attractiveness of projects (as can be concluded from the previous analysis), with accompanying diminished petroleum-company activity. However, exactly the contrary occurs when the companies presume that the tax rate increase is only temporary. Exploration investments can be well profitable if they can be treated as direct write-offs for income-tax purposes. Exploration is in this case temporarily less risky because the tax credit on exploratory activity is high.

This single example demonstrates that the effect on the petroleum industry of temporary changes in tax rates is difficult to estimate. The topic of the level of the corporate-income-tax rate in relation to the economy is, however, far beyond the scope of this book.

Changes due to specific circumstances

Renegotiation desired by the companies

The position of the coal industry in western Europe is a clear example of conditions arising which necessitate radical changes in mining legislation to meet the industry's demands. In western Europe the coal industry is declining due to technical and structural economic conditions. Since the industry played an important role in the economies of a large number of European countries, special arrangements were necessary to prevent economic chaos and to promote a gradual and socially-acceptable decrease in activities. Large subsidies were offered by most European governments to cope with the problems, and exemptions from several payments due to the government were obtained by the companies.

Similar conditions may well prevail within a few decades in the petroleum industry as other energy sources, such as nuclear energy, become competitive and eliminate the petroleum industry from most of its present markets.

Frequently, prices established in contracts, are renegotiated as soon as conditions change. In the U.S.A., for example, companies have frequently expressed the desirability of renegotiating the price levels for gas deliveries.

Renegotiation desired by governments

The rapid changes in the economic, technical and political conditions in the petroleum industry make frequent renegotiation of contracts advisable. The post-World War II period displayed rapid political developments. The key process was the decolonization of the developing countries. This resulted in renegotiation of a number of petroleum contracts originating in the colonial or semi-colonial period. As is illustrated in Chaper III, such changes in the petroleum law occasionally resulted in grave political troubles as in Iran from 1951 to 1954. Most O.P.E.C.-countries feel that a number of contracts are still outdated and need renegotiation.

Besides profound political changes, the economic and technical conditions in the petroleum industry are changing continuously. In the long run a tendency exists for costs to decrease, as illustrated in Chapter III. This tendency must ultimately result in a renegotiation of most contracts, since in the very long run it means that the relation between costs and revenues upon which the original contracts were established will be fundamentally altered. This is especially true of natural gas in the developing countries.

Finally, renegotiation may be called for when unexpectedly large profits accrue to the company, amounting to excessively more than the quasi rent on the projects. The government cannot tolerate continuation of this unforeseen state of affairs because of the extreme detrimental effects on the country's economy.

For instance, the discovery of the Slochteren gas field completely altered the petroleum economy of The Netherlands; excessive rents would have been earned by the operating companies if the legislation applicable to former oil and gas fields had been retained. The old Mining Code of Napoleon implicitly provided for the possibility of renegotation after a discovery, because the grant of a special production permit was required for each individual case. For this reason the Dutch government was able to "renegotiate" the terms.

Similar conditions prevail in the Middle East, where post-war developments completely altered the area's petroleum economy and significantly enlarged the expected rent on the projects. At present, renegotiation of most petroleum contracts would prove beneficial to these governments and the societies they represent.

The continuously shifting pattern of the world-petroleum economy, technology and politics make it expedient for renegotiation clauses to be included in all petroleum legislation which grants concessions over several decades.

An extreme alteration in mining legislation provides for the *nationalization* of mineral properties. This is a politically-loaded problem. From certain ideological standpoints (socialistic or nationalistic) nationalization may be advantageous to the state even when the enlargement of the proceeds in this way is questionable. The discussion of this facet of nationalization is beyond the scope of this study.

Insofar as nationalization is regarded by governments as a means of enlarging proceeds from mining operations, it must be examined here. The proceeds that are earned by the government from the nationalization depend entirely upon its chronological occurrence. Nationalization of a property at the exact beginning of the project (an extreme example) is not a nationalization but the founding of a state company. A nationalization at the end of the lifetime of a project has no relevance to the proceeds of either the private company or the government.

A "true" nationalization occurs during the lifetime of a project, normally when the mineral production has proved itself to be profitable. The longer a government waits with nationalization, the less the added proceeds that can be expected. The government generally compensates the private company for the loss of property.

A government is successful with nationalization when it earns with its move the entire "true rent" which the project is capable of generating. Of course, the government can earn even more if the private company is not compensated fully for the loss of its property. In this case the effects of the project on the private company have been less than acceptable.

Nationalization usually changes the economic environment of the mining project. It is not certain, therefore, that the government is earning the original true rent through nationalization, even when the timing was optimal. For instance, if the company's management is replaced by a less capable group (because the former group is withdrawn by the private company), investments per bbl and the cash outlays per bbl may become larger due to the untrained personnel. Even when trained personnel is on hand to refill the vacated positions, certain costs may become higher because several contracts must be revised. Revenues per bbl may be lower following the loss of transport facilities of traditional markets. The final result may be that the proceeds gained from nationalization may be lower than the proceeds that would have been earned with the original mining legislation.

Even if the nationalization of one project succeeds for the government (resulting in more proceeds than would be obtained in other cases) exploration activity of private companies in the area may cease for fear of further nationalization.

Consequently, a government that is planning nationalization of oil properties within its borders should remember that the result may be a decrease rather than an increase in the proceeds of mining activities. However, especially if large rents were flowing out of the country, and if the move is carefully planned, such a nationalization can be beneficial to the state.

A nationalization of the petroleum industry within a country usually leads to extensive political troubles between the country and the company. The company will frequently accuse the government of illegal actions, and the company's mother-country may become involved in the conflict on the side of the company. This was recently the case with the nationalization of the petroleum industry in Peru, where private business obtained considerable support for its viewpoint from the U.S.A.

Since nationalization is a drastic change in mining legislation generally not included in "renegotiation" clauses, it must stem from a major political decision on the part of the government. In the process of decision-making, the government must balance the positive aspects of a large national share in the industry and the earning of increased proceeds against the negative aspects of the possibilities of damaging political relations and economic conditions.

The goal of a larger national share in the petroleum industry can best be achieved by moving along the path of negotiations with the companies to secure economic and political stability. Such negotiations, however, are only useful if the companies concede that the state must earn the true rent of the projects.

Conclusions

From the previous discussion it can be concluded that a number of conditions must prevail to maximize public revenue from mining activities. These conditions are presented in this chapter under the heading "Conditions to achieve the maximum public revenue". These conditions, however, embody a somewhat static approach to the problem. Therefore, the inclusion of a renegotiation clause is needed in each petroleum law to cope with the continuously changing technical, economic and political conditions.

The maximum public revenue, however, is a partial goal. The activities of the petroleum companies contribute more benefits to the state than merely direct payments to the government. These benefits must be included in the studies preceding the preparation of a final draft of a petroleum law.

Literature

Anonymous, 1970. Capital outlays sharply higher. *Petrol. Press Serv.*, 37(2):57-59

Arps, J. J., and Roberts, T. G., 1958. Economics of drilling for Cretaceous oil on east flank of Denver Julesbury basin. *Bull. Am. Assoc. Petrol. Geologists*, 42(11): 2549-2566.

Arps, J. J., Brons, F., van Everdingen, A. F., Buchwald, R. W., and Smith, A. E., 1967. A statistical study of recovery efficiency. *Am. Petrol. Inst. Bull.*, D 14: 1-15

Bradley, P. G., 1967. The Economics of Crude Petroleum Production. In: J. Johnston, J. Sandee, R. H. Strotz, J. Tinbergen, and P. J. Verdoorn (Editors), *Contributions to economic analysis*. North Holland Publ. Co., Amsterdam, 149 pp.

Bijl, P. C. J., 1968. Surface installations and operations. In: KONINLIJK NEDERLANDS GEOLOGISCH MIJNBOUWKUNDIG GENOOTSCHAP (Editor), *Symposion on the Groningen Gas Field—Verhandel. Koninkl. Ned. Geol. Mijnbouwk. Genoot., Geol. Ser.,* 25: 59-65

Cameron, B. A., 1966. The petroleum prospects under the marine areas in the world. In: P. HEPPLE (Editor), *Petroleum Supply and Demand.* Elsevier, Amsterdam, pp. 23-60.

De Wolff, P., 1964. Economische aspecten van het aardgas. Overdruk *Academiedagen,* dl. XVI, Centraal Planbureau, 's Gravenhage, pp. 61-84.

Fisher, F. M., 1964. *Supply and Costs in the U.S. Petroleum Industry—Two Econometric Studies.* Johns Hopkins, Univ. Press, Baltimore, Md., 178 pp.

Hodges, J. E., and Steele, H. B., 1959. An investigation of the problems of cost determination for the discovery, development, and production of liquid hydrocarbons and natural gas resources. *Rice Institute Pamphlet,* 16(3): 4-153.

Levorsen, A. I., 1967. *Geology of Petroleum.* Freeman, San Francisco, Calif., 724 pp.

Lovejoy, W. F.; Homan, P. T.; Galvin, C. O., 1963. Cost analysis in the petroleum industry. *J. Graduate Research Centre, Southern Methodist Univ., Dallas, Texas,* 31(1 and 2): 105 pp.

Smith, C. A., and Brock, H. R., 1959. *Accounting for Oil and Gas Producers—Principles, Procedures and Controls.* Prentice Hall, Englewood Cliffs, N. J., 536 pp.

Taher, A. H., 1966. *Income Determination in the International Petroleum Industry.* Univ. Calif. Press, Berkely, Calif., 227 pp.

Chapter VII

The Dutch offshore-mining legislation

Introduction

The previous paragraph produced a set of conclusions that should be applied to extract the maximum public revenue for the government from mineral operations. It is interesting to check these conclusions against some actual conditions. One of the most interesting developments in the field of mining legislation offshore has taken place in The Netherlands. This history can be profitably followed as a first control. In the next chapters the developments in more countries in the world will be analyzed.

The developments in The Netherlands were particularly interesting because the legislation passed through four governments of different political settings before it was completely accepted by the parliament. Further, this development took place after the discovery of the giant Slochteren gas field. This field changed the European energy situation considerably and it also influenced the interest of the oil companies in the continental shelf beneath the North Sea.

The discovery date of the Slochteren gas field was August 14, 1959. About three years were necessary to deliniate the field. A giant gas field was discovered as a result of this research. The developments in the Slochteren area drew the attention of the world towards the North Sea as a possible giant source of gas and oil—an area that lies directly in the middle of the very important western European market. In 1962 the N.A.M. (Shell/Esso) drilled a few exploratory wells not too far from the Dutch coast in the North Sea. The results were only dry holes, but they could not curb the enthusiasm for the North Sea exploration area. The Dutch government, afraid for chaotic exploration, declared that all the drilling activities in the Dutch offshore waters would be regarded as an unfriendly action against the Dutch State. In the meantime the government (being the government-Marijnen with a liberal* character) prepared a concept for offshore-mining legislation. The right to develop this legislation stemmed from the Treaty on the Continental Shelf concluded at Geneva on the 29th of April 1958.

The first draft of the Dutch mining act was presented to the Parliament on the 9th of June 1964—a proposal of a very general character. Only the possibilities of raising bonuses, etc. were established. The exact figures for these elements were to be given in a separate decree. This would make the mining legislation more flexible, since a change in the exact figures for the components of the bill could be handled by a simple decree that would not need to pass the parliament. The government-Marijnen did not succeed in obtaining acceptance for this act, because it fell before the act passed the Parliament. The new government, the government-Cals, had a partly socialistic background. It continued

---

*Note that the European "liberal" is not the American one. In fact this type of liberal resembles more closely the American "conservative".

the work of the previous government but slightly changed the character of the mining act by extending the possibilities for state participation. This Act—Continental Shelf, Mining Act ("CSMA")—was accepted the 23 rd of September 1965. This act will be discussed in the following paragraph more closely.

As soon as the act was accepted the government-Cals began the preparation of the decree on the exploration and production licences to specify the bonuses, surface rights, royalties etc. This government, however, fell in the autumn of 1966, before it could handle this decree. The succeeding government-Zijlstra was again of a more liberal character and changed the prepared version on several points. This decree was proposed to the parliament at the 27th of January 1967 and the parliament decided to handle it as a separate Act. This act was accepted on the 2nd of August 1967, under the fourth government-De Jong. This government did not change the content of the prepared decree. It was interesting, however, that the Parliament was informed about the changes that the government-Zijlstra brought into the still unpublished proposal of the government-Cals. This makes it possible to study two different versions of the decree: a more socialistic one and a more liberal one. These two versions will be compared and evaluated in this chapter.

The result of the apparently endless discussions about the Dutch offshore-mining legislation was that drilling was stopped for a good three years on the Dutch side of the continental shelf. In the German, Norwegian, Danish and British areas, however, an active exploration was going on, as can be seen in Fig. 39. The results in the German, Danish, and Norwegian area were disappointing because 17 wells were drilled in this area towards the end of 1966 without a single commercial discovery. On the British side, however, the results were very promising. Various prolific gas fields were discovered in these years. For the North Sea continental shelf area as a whole the number of discoveries of commercial gas or oil fields was about one out of four exploration wells. This rather successful figure was reached by the end of 1966 and remained unchanged during 1967 as can be seen on Fig. 39.

The preparation and debate over the decree on the exploration and production licences took place in a situation where the realized success ratio of the exploration in the non-Dutch parts of the continental shelf of the North Sea was about one producer out of four exploratory wells.

In the meantime, the exploration on land in The Netherlands continued during the preparation of the offshore-mining legislation until the 14th of December 1965. In 1965 the on-land exploration had proved highly successful, with 16 out of 40 exploration wells revealing commercial gas wells. Since in The Netherlands the Mining Law of 1810 was applicable on land—a law totally unsuited to efficient exploration—operations had to be stopped while awaiting preparation of new legislation for land. The good results until the end of 1965, however, were an indication of geological allure of the Dutch offshore. Thus the attractiveness of the Dutch offshore was enhanced by good results on land and the apparently facorable situation on the British side of the North Sea continental shelf. Most companies probably estimated in 1967 the probability-of-success to be better than one commercial discovery out of four exploration wells. This statement, however, cannot be proved.

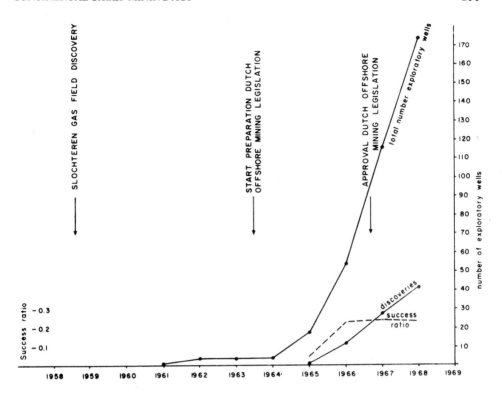

Fig. 39. Exploratory results in the North Sea area

The following sections of this chapter will handle the Continental-Shelf-Mining Act and the Decree on the exploration and production licences. In these sections the assumption will be made that the government had to estimate the success ratio as about one commercial producer out of four exploration wells due to the developments offshore and onshore in the years 1965 and 1966.

Continental-Shelf-Mining Act

The official name of this act is: "Act of 23 September 1965 embodying regulations with regard to the exploration for and the production of minerals in or on the part of the Continental Shelf that is situated underneath the North Sea". It is very general in content and goes into no detail—as will soon become clear. The total act comprises 43 articles of which the majority are not relevant to this section's discussion because they do not apply to the problems dealt with in this book. Only the relevant articles will be discussed.

Selected articles from the Continental-Shelf-Mining Act

Article 3

The text of Article 3 is as follows:

"It can be provided for by general administrative order that for parts of the Continental Shelf which are thereby indicated, no licences or exemptions shall be granted".

This article makes it possible to close a certain area to exploration and production. It is important because it can be used in the case of a very high geological risk. In high-risk situations, as was concluded in the previous chapter, it is possible to permit exploration in a limited area to analyze the geological results so that a possible new legislation can be better adapted to a more closely calculated probability-of-success. The article does not appear to have been included for this purpose, however. The reason for closing certain areas to exploration is the possible danger for sea transport or military operations in such areas.

Article 8

The text is: "To a licence the regulation can be attached that the holder will owe the State, at the points of time as laid down in it, area-duties which have been laid down in it".

This article opens the possibility of raising-annual-fixed or rising-surface duties. As can be seen, the text of this article is very vague as is the case with all the articles in this general act.

Article 10.3

The text of the third part of Article 10 begins: "moreover the following regulations can be attached to an exploration licence or a production licence:
a.   that the holder will be indepted to the State, when the licence becomes effective, an amount to be fixed therein;
b.   etc."

In this part of Article 10 the possibility is established for the State to demand a bonus at the beginning of the exploration or production. It can be concluded from the text that an initial bonus is required as soon as the licence is granted. No bonus is required at the discovery date or any other date after the start of the exploration. There simply stands that a bonus must be paid directly on or by the date when the production licence becomes effective.

Article 11

Article 11 is probably the most important and controversial article of this legislation. The text follows·

"1. To a production licence the regulations can be attached that the holder owes the State:
a.   at specific points of time, as laid down in the licence a fixed amount, as laid down in it, to be calculated in accordance with the value of the minerals produced;
b.   at specific points of time, as laid down in the licence a fixed amount, as laid down in it, to be calculated on the profit made as a result of the production.

2. To a production licence for a mineral can furthermore be attached the regulations that the holder, if he has proven by utilizing that licence or an exploration licence, the existence of such mineral in a quantity that can be economically produced:

a.   will cooperate as required by Our Minister in establishing a limited liability company (corporation) for the production of that mineral, in which the holder and the State will take an interest in a manner to be laid down in the licence, as well as in the establishment of an agreement between the holder and such limited liability company (corporation) in virtue of which agreement the holder will produce exclusively for the account of said limited liability company (corporation).

b.   etc."

This text implies the possibilities of imposing royalties, tax, and of requiring state participation. Article 11.1.a. opens the option of demanding fixed royalties or royalties according to a sliding scale. Article 11.1.b. is included in the legislation to make it possible for the government to raise a tax similar to the corporate-income tax. As soon as the company locates an economically-recoverable reserve of minerals this company must be willing to accept state participation. The royalties, the tax and the state participation are described in rather vague termes. A striking feature of this article is the potentiality of State Participation. State Participation becomes possible when an economically-recoverable reserve is discovered; by this law the state escapes the geological risk that is facing the mining company. This form of state participation might influence the exploration activity in a negative way as is described in Chapter V. Further, the recognition of an economically-recoverable reserve in the early days of the discovery of a mineral deposit, is no easy matter. There remain a number of geological risks to be taken even after the discovery, especially during the deliniation of the field—before it can be concluded whether or not the reserve is economically recoverable. This problem is particularly important if a marginal or unimportant reserve is discovered. In this case it is often difficult to conclude whether the post-discovery investments are worthwhile. Much criticism was passed upon this fact in the Parliament.

Article 18.1.

The last article to be considered here is the first part of Article 18. The text reads as follows: "If the holder of a licence or of an exemption asks for an alteration by letter, Our Minister—with due observation of the stipulations laid down by or in virtue of this law—will grant such a request while making the necessary further alterations, if any, in so far as it can be considered compatible with the general interest".

This article provides the possibility for changes in the licence (or exemption) in the case of special circumstances. This possibility was necessary considering the rather heavy burden which mining legislation can impose on the companies. This article, however is not applicable to oil and gas.

The general character of the Continental-Shelf-Mining Act

The Continental-Shelf-Mining Act is a general act applicable to all minerals eventually produced from the continental shelf. Specific regulations for the exploration and production of oil and gas must be handled in separate decrees. Two important decrees are the general administrative order concerning exploration and production licences and the mining regulations, continental shelf. The former decree will be handled in this chapter.

The general character of this act agrees with the conclusions of Chapter VI. Article 18.1 provides the means for a separate treatment under special conditions. The possibility for raising bonuses, surface duties, royalties, and tax and for the application of state participation is presented. The decree on exploration and production licences, however, must be handled before any definite conclusion can be drawn.

An interesting aspect of the Continental-Shelf-Mining Act is that a distinction was made between an exploration and a production licence. The exploration licence was defined as: "a licence granted for the purpose of carrying out an exploration survey into the minerals mentioned therein, as well as to carry out a reconnaissance survey". The production licence is a licence granted for the purpose of producing minerals mentioned therein, as well as for conducting an exploration survey into such minerals and for conducting a reconnaissance survey. This distinction is interesting because in most oil legislation the right to explore for petroleum usually implies the right to produce the petroleum when it is discovered. The distinction between these two types of licences was practical because the government must decide about participation after the presence of an economically producible quantity of minerals has been found. Since the exact terms of the state participation must be negotiated, the split into two different licences seemed justified. The government declared, however, that the production licence would normally be granted to the company that discovered the minerals.

Decree of 27 January 1967

The government-Cals prepared this decree, but was unable to present it to the parliament before it fell in the autumn of 1966. The government-Zijlstra presented its own version of the decree on the 27th of January to the Dutch Parliament. As was explained in the beginning of this chapter, two versions of this decree are known: a socialistic version (of the government-Cals) and a liberal one (of the government-Zijlstra). Since the last was accepted, this version will be discussed in this section.

This decree consists of eleven articles. These articles are numbered with Roman numerals and subdivided into articles numbered with arabic numerals. For this discussion some parts of the Article II, III, and IV are of importance. Article II specifies the restrictions and provisions related to the exploration licence. Article III handles the production licence, while Article IV describes in more detail the regulations for state participation.

In Article I a natural-gas deposit is defined, since different regulations occur in some articles for oil and gas. This definition is such that only a deposit of dry and non-associated gas is considered "natural-gas deposits".

The exploration licence (Article II)

Bonus. The bonus is described in article 4: "Starting from the day after the one upon which the licence has become effective, the licensee owes the State a sum as referred to in

Article 10, section 3 of the Act (bonus) in the amount of f 1000 per full square kilometer, calculated according to the area of the licence area".

Since the surface of most of the blocs is about 400 square kilometers (156 square miles statute), the bonus to be paid per bloc was about f 400,000.– or U.S.$ 110,000.–. This corresponds with a bonus of about U.S.$ 710.– per square mile, or somewhat more than one U.S. dollar per acre. This initial bonus is extremely low compared with bonuses paid, for instance, in the U.S.A. Sealed bidding sometimes produces bids of U.S.$ 685.– per acre (offshore, Louisiana, 1967) or even U.S.$ 1,600.– per acre (St. Barbara Channel, California, 1968). These highly prolific areas are not entirely comparable to the Dutch offshore, but the examples illustrate clearly that a bonus of one U.S. dollar per acre is low and consequently favorable for the companies.

Annual-surface duties. The annual surface-duties are described in article 5: "Starting on the day after the date upon which the licence has become effective, the licensee annually owes the State a surface duty, as referred to in Article 8 of the Act, calculated according to the area of the licence area on that day of the year concerned on the basis of the following scale, wherein the latest wage index figure, prior to that day of the year, indicated in pursuance of Article 9, tenth section, of the Compulsory Old Age Insurance Scheme (Statute Book 1956, 281), is specified with the letter a, and the latest wage index figure in pursuance of the aforementioned statutory provision before the point of time upon which the Continental-Shelf-Mining Act becomes effective is specified with the letter b.:

a/b x f 50 per full km^2 for each of the first five years

a/b x f 100 per full km^2 for each of the successive five years

a/b x f 150 per full km^2 for each year of the remaining currency of the licence."

The surface duties are rising ones. The first rise of f 50 per full square kilometers occurs after five years, and the second rise after ten years. The rising-surface duties are, however, even at their highest level not seventeen (U.S.dollar) cents per acre. These annual-surface duties are consequently low and they favor the companies.

General. The low amounts to be paid for the bonus and the annual-surface duties indicate that the terms of the exploration licence are highly advantageous for the companies. These favorable terms clearly stimulate exploration activity.

The Dutch conditions for the exploration licence are in agreement with the conclusion (Chapter VI) that the weight in a mining act should not be given to bonuses and surface duties, since such provisions can engender considerable losses of economically-recoverable reserves when heavily applied; and because they can have a heavy negative influence on exploration activity because of the pre-discovery character of their payment.

The terms of the exploration licences were so favorable that it was feared that several companies would begin speculating on blocs. These terms made it appear profitable to buy the blocs outright, to hold them without exploring them, and to sell them when the geological conditions proved beneficial. This type of speculation is, of course, not in line with the interest of the Dutch State; consequently a reasonable work obligation was included in the exploration licence. This obligation assured that the licensee be required to commence reconnaissance and exploration investigations in the licence area and subsequently to energetically continue the investigations. During the first five years the licensee must spend about U.S.$ 7.— per acre on exploration and reconnaissance. This amount will be easily spent if a normal program is followed, but the working obligation is large enough to avoid misuse of the concession.

The production licence (Article III)

Annual-surface duties. Annual-surface duties are also required in the production licence. The first sub-article of Article III orders an annual-surface duty of a/b x f 300 per full km². In this article the meaning of the letters a and b is the same as in Article II—5. The surface duties are higher for the production licence than for the exploration licence. The government's aim with these surface duties is to stimulate the companies to reduce the production area voluntarily as much as possible. In general, however, it can be stated that a surface duty of something more than U.S.$ 0.30 per acre is not particularly high.

Royalties for oil and natural-gas liquids. The royalties for oil and natural-gas liquids are fixed in article 3 (of Article III):
"1. The licensee annually owes to the State an amount as referred to in Article 11, first section under a. of the Act, consisting of a percentage of the total value of any petro-leum* which has been produced under the licence, but not for account of a limited liability company, established in accordance with article 2 of Chapter IV or article 2 of Chapter V and which has been removed from the licence area in the previous calendar year.
2. The percentage mentioned in the first section is fixed on the basis of the following scale:" (see Table XV).

This article introduces the sliding-scale royalty into the production licence. The observations of Chapters IV and V will be discussed under the heading "state participa-tion". The sliding-scale is such that when a large oil field is in production—for instance producing 100,000 bbls per day (about 6 million m³ per year)—the weighted average of the royalties to be paid is only about 12 percent. The sliding scale never exceeds 16 percent. Since fields producing 100,000 bbls per day are exeptions, the weighted average of the royalties will be in general less than 12 percent. This percentage is in accordance with or lower than royalties required in other countries legislation as will be shown in the next chapter.

It is essential that the royalty does not weigh too heavily in the range of low outputs, since it is in this range that marginal fields have the greatest chance of occurring. Up to an output of about 10,000 bbl per day, the royalties to be paid under this production licence are almost negligible (up to 3 %). The essential function of the sliding scale—the avoidance of sub-marginal projects that could be profitable without royalty—is realized with this scale.

Two questions arise when the royalty is introduced into the legislation. The first is: "What is the total value of the petroleum?" The second is: "What is the influence of the

---

* In the translation of the texts of the various articles petroleum stands for crude oil and natural-gas liquids.

TABLE XV

SLIDING SCALE FOR ROYALTIES APPLICABLE TO OIL AND APPLIED IN THE DUTCH OFFSHORE-PETROLEUM LAW

For the portion of the annual removal of:	the following percentage applies:
0 to 100,000 m^3	0
100,000 to 200,000 m^3	2
200,000 to 300,000 m^3	3
300,000 to 400,000 m^3	4
400,000 to 500,000 m^3	5
500,000 to 600,000 m^3	6
600,000 to 700,000 m^3	7
700,000 to 800,000 m^3	8
800,000 to 900,000 m^3	9
900,000 to 1 million m^3	10
1 million to 2 million m^3	11
2 million to 3 million m^3	12
3 million to 4 million m^3	13
4 million to 5 million m	14
5 million to 6 million m	15
more than 6 million m^3	16

areal extent of the licence area on the amount that can be removed from that area?" Both questions merit a closer examination.

The value of the petroleum is in principle determined by its sales price. This price may be reduced by duties and taxes levied by any State for the import and supply of the petroleum, by the cost of processing and by transportation costs. The latter two costs are subject to some restrictions. The sales price may be used as long as it is higher than the average price at which oil of an equal quality normally has been imported in the countries bordering on the North Sea, during the calendar year concerned. Consequently the royalty is calculated as a normal wellhead price, which must be in accordance with the local oil market.

The areal extent of the licence may be such that an entire oil field is included in the licence. Even if the oil field is spread beneath several blocs (of 156 square miles), the licence may cover the same area, because a licence may include several blocs. It is, however, logical that the larger the productive area of the oil field, the larger the chance that several companies will produce from the same field, out of different licences. In this case the royalty paid may equal less than ten percent, even though the field produces 100,000 bbls per day, because the field is produced from several licences.

Royalties for natural gas. The royalties for natural gas are given in article 6:
"1. The licensee owes annually to the State an amount as referred to in Article 11, first section under a. of the Act, consisting of a percentage of the total value of any natural gas produced under the licence and removed from the licence area during the previous calendar year.
2. For the application of the first section, removal from the licence area does not include

the removal from the licence area of natural gas, which is pumped elsewhere into the continental shelf for the purpose of the exploitation of petroleum deposits.

3. The percentage mentioned in the first section is fixed on the basis of the following scale" (see Table XVI).

TABLE XVI

SLIDING SCALE FOR ROYALTIES APPLICABLE TO NATURAL GAS, APPLIED IN THE DUTCH OFFSHORE-PETROLEUM LAW

For the portion of the annual removal — with a pressure of 1 atmosphere and a temperature of 15 degrees Celsius — of:	*the following percentage applies:*
0 to 100 million m^3	0
100 million to 200 million m^3	1
200 million to 300 million m^3	2
300 million to 400 million m^3	3
400 million to 500 million m^3	4
500 million to 600 million m^3	5
600 million to 700 million m^3	6
700 million to 800 million m^3	7
800 million to 900 million m^3	8
900 million to 1 billion m^3	9
1 billion to 2 billion m^3	10
2 billion to 3 billion m^3	12
3 billion to 4 billion m^3	14
4 billion to 5 billion m^3	15
more than 5 billion	16

Gas production is also subject to a sliding-scale royalty, with exception of that gas which is used elsewhere on the Dutch part of the continental shelf. The same remark that can be made for oil applies for gas: the weighted average of the royalties to be paid for even significant production is not high. For instance, the average royalty for annual production of 100×10^9 cuft. (3×10^9 m^3) is less than 9%. Only in situations where a field can be found with an annual production of 10^{12} cuft., as is expected from the Slochteren gas field within a few years, will the royalty amount to nearly 16%.

The price of gas is calculated in about the same way as that for oil. Duties and taxes, transport costs and treatment costs may be subtracted from the sales prices, as is explained in article 7 of the decree. The royalty is consequently based on the wellhead price. In the case of natural gas the price of the gas must be approved by the government and is not related to the market.

The same observation concerning the relation between the areal extent of the licence area and the amount that can be removed from it can also be applied to natural gas. For instance, the Slochteren gas field would cover more than one bloc if it had been found beneath the North Sea at the same depth, because it covers about 290 square miles (750 km^2) in the province of Groningen. If such a field were found beneath the North Sea, and was situated under several licences, a production of 10^{12} cuft. out of one licence would be improbable.

Taxation. The articles 12 through 21 provide the essential requirements for the payment of taxes similar to the normal corporate-income tax raised by the Dutch State. Some of the essential sentences are quoted here:

"article 12. 1. The licensee annually owes the State an amount as referred to in Article 11, first section, under b. of the Act, of 50% of the surplus balance of a profit and loss account in compliance with articles 13 and 14 over the expired financial year, which includes the costs and returns of the reconnaissance, exploration and production of minerals in pursuance of the licence relating to that year.

2. A financial year . . . etc.

article 13. 1. In making up a profit and loss account as mentioned in Article 12, the credit side is attributed with:

a. the returns from the minerals produced in pursuance of the licence;

b. etc.

2. In making up a profit and loss account as mentioned in article 12, the debit side is attributed with:

a. the costs of conducting reconnaissance and exploration investigations incurred after the licence has been granted and the costs of producing and supplying the minerals insofar as the direct as well as any reasonable indirect and general costs are concerned;

b. the costs of reconnaissance and exploration activities other than those mentioned under a. incurred after the licence has been granted, insofar as the licensee carries out these activities via an undertaking of a permanent character in The Netherlands, with the exception of those costs, which have already been charged to another profit and loss account;

c. the following depreciations:

costs not already charged to another profit and loss account, incurred before the licence has been granted, insofar the licensee has performed the reconnaissance and exploration surveys concerned via an undertaking of a permanent character in The Netherlands, up to an amount of 1.00 guilder—10%; on any permanent productive resources up to an amount of 1.00 guilder on the basis of the estimated length of life;

d. the taxes and other Netherlands statutory charges owing to the State insofar as they can be considered as company charges for activities mentioned under a. or b., with the exception of the taxes and charges, which already have been charged to another profit and loss account, the taxes levied on income, profit or capital and their advance levies as well as amounts owing pursuant to the licence, calculated on the profits made from the production;

e. an amount of 10% of the costs mentioned under a—c.

3. etc.

article 14. If, during any financial year, a profit and loss account shows an adverse balance, it is transferred to the debit side of the profit and loss account of the following financial year"

A number of important remarks can be made concerning these three articles. The tax to be paid is 50% of the yearly profit calculated according to the articles 13 and 14. This tax is not excessively high.

More important than the exact percentage, however, is the way in which the yearly profit can be calculated. From article 13 it can be seen that not only direct but also indirect and general costs can be attributed to the debit side. This is important because the oil and gas industries are more and more distinguished by large indirect and general costs. As soon as the production licence has been obtained, all the subsequent exploration and reconnaissance-survey costs can also be expensed directly from the yearly profit. Surveys outside the licence-area can be deducted as long as these are surveys to study the Dutch part of the continental plat. Exploration and reconnaissance costs made prior to the date that the licence is granted, can be attributed to the debit side as well, but these costs must be depreciated.

An interesting aspect of this taxation is the possibility of subtracting 10% of the majority of costs from the yearly profit. This percentage makes the "offshore taxation" almost equal to the normal-corporate-income tax in The Netherlands, which amounts to 42% of the yearly-taxable profit. This "investment credit" of 10% clearly stimulates exploration activity, especially since the costs for exploration and reconnaissance prior to the date that the production licence was granted, and the costs of activities outside the licence area on the Dutch continental platform are included in the calculation for this investment credit.

Another positive aspect of this regulation is the possibility of transferring the losses of previous years to the next fiscal year. The point in time where taxes must be paid can be delayed in this way almost until the pay-out time is reached, because there is no time limit for this transfer of losses.

State participation (Article IV)

State participation is described in Article IV. It reads: "If the holder of an exploration licence under that licence has proven in a natural gas deposit the presence of natural gas exclusively or not exclusively, in an economically producible quantity, and consequently, a production licence is granted to him by virtue of Article 13, first section of the Act and thereby the right is exercised as referred to in Article 11, second section under a. of the Act, such a production licence, furthermore, cannot be granted, but with the following provisions.

article 1. For the application of article 2, a natural gas deposit is understood to mean: a deposit of bitumens, in which no petroleum is found in addition to the natural gas found in it, which natural gas is capable of being independently produced in a normally conducted production enterprise, and in which deposit natural gas is found, which when produced is so deficient in natural gasoline that the development of the deposit, in a normally conducted production enterprise, is aimed at the production of natural gas.

article 2. The licensee is required to cooperate in:

a. the establishment with due observance of articles 3 and 4, of a limited liability company for the production under the licence of natural gas or petroleum from each natural gas deposit, in which company the licensee participates for 60% and the State or the limited liability company appointed in the licence participates for 40%, and

b. the conclusion of an agreement between the licensee and the limited liability company to be established, in pursuance of which the licensee will produce only for the account of that limited liability company, and which agreement contains the provisions referred to in article 6.

article 3., etc.

article 4.1., etc.

2. In the memorandum of association it will further be provided that the licensee immediately will be compensated by the other shareholder to 40% of the amount referred to in the first section."

Typical of the state participation on the Dutch offshore-mining legislation is that participation is restricted to the production of "dry" and non-associated gas and that state participation is 40%. The occurrence of "dry", and non-associated gas is highly likely because of the important "dry" discoveries that have already been made onshore in The Netherlands and offshore on the British side of the continental platform (including the Slochteren and the Leman-Bank fields).

State participation is restricted to 40%, but the statutes of the limited-liability company include the provision that the approval of 2/3 of the votes of the general meeting of shareholders is necessary for most important decisions, such as the annual investment and financing plan, the allocation of the profits, etc. These provisions are given in article 3 in more detail. Consequently, the participation of 40% is sufficient to influence most of the major decisions of the company. It is significant that the licensee is compensated immediately for the 40% of the costs that he invested in the project. This compensation is still something of a burden to the licensee, since, while he is returned 40% of the costs incurred prior to state participation, he is not compensated for the interest on these costs. As explained in Chapter V, this method of handling the compensation diminishes in general the attractiveness of the project.

Article V regulates state participation when a dry natural-gas-production licence is necessary in situations not included under Article IV.

Note that the royalties to be paid are diminished by exactly half when state participation occurs. The sliding scale is the same as for a gas-production licence without state paricipation, but the maximum percentage is 8%. This is a royalty-credit designed to meet the negative effect of the state participation. This royalty-credit is somewhat important, because for large productions—the cases in which state participation is likely—the credit is nearly a weighted-average royalty of 8%.

General remarks concerning the production licence with or without state participation

The terms of the production licence are such that it must be assumed that the maximum public revenue will be reached, according to the conclusions of Chapter VI.

Different regulations have been structured for oil and for gas. The accent of the decree is on sliding-scale royalties, tax, and state participation.

If we assume that the government estimated the probability-of-success to be 0.25, the conditions pressure for oil seems rather low; for gas it is about as was concluded in Chapter VI.

Since surface duties and bonuses play a minor role, the conditions pressure is determined by the sliding-scale royalty, the tax and the state participation.

For oil, the tax will be 50% but an important investment credit of 10% is granted, while the sliding-scale royalty may be up to 16%. No state participation is required. These conditions are well suited to a generally-expected-moderate probability-of-success $(0.1 < p < 0.25)$. The conditions pressure for oil is consequently somewhat low.

For natural gas, the tax of 50% with the investment-credit of 10% is equally applicable. The sliding-scale royalty is up to 8% when state participation is applied. For an economically-recoverable reserve a state participation of 40% is possible. These conditions also fall within the range of the generally-expected-moderate probability-of-success, but the conditions pressure will be higher for gas than for oil—at least for the reasonably profitable projects—and consequently well adapted to reaching the maximum public revenue.

The general character of the Decree of 27 January 1967

From the previous discussion it can be concluded that the terms for the exploration licence are such that exploration is stimulated. The terms of the production licence are for gas in accordance with the conclusions of Chapter VI, and for oil somewhat less so. The government is aiming at high ultimate earnings by laying stress on the royalties, tax, and state participation instead of high bonuses and surface duties. In this way the government is facing geological risk of limited proceeds if exploration proves unsuccessful.

The rather low conditions pressure for oil can probably be explained by the fact that western Europe is a major importer of crude oil and that the government—Zijlstra wanted to stimulate the exploration for oil as much as possible.

Decree, as prepared by the government-Cals

The government-Cals prepared the decree for the exploration and production licences. Since this government fell before the decree could be approved in the Parliament, it has never been accepted in its entirety. The government—Zijlstra changed some essentials before presenting the document to the Parliament. The version of the government—Cals, a socialist one, was published by the government—De Jong in May 1967. This version will be discussed in the section which follows.

The differences between the version—Cals and the version—Zijlstra are restricted to the sliding scale used for the royalties, the state participation and the taxation.

Sliding scale

The sliding scale for oil and natural-gas liquids was different in the original draft from that in the Decree of 27 January. The scale is given in Table XVII. As can be seen in this table, the scale differs only for very large yearly productions. The average royalty on a concession producing 100,000 bbls per day (6×10^6 m^3 per year) is 11.0%, which is almost the same as the 11.7% royalty according to the Decree of 27 January. For

TABLE XVII

SLIDING SCALES FOR ROYALTIES TO BE PAID ON OIL AND GAS PRODUCTION, AS PRESENT IN THE NOT-ACCEPTED VERSION OF THE GOVERNMENT-CALS FOR THE DUTCH OFFSHORE-PETROLEUM LAW

Oil		Gas	
production 10^6 m³/year	perc.	production 10^9 m³/year	perc.
0 − 0.1	0	0 − 0.25	0
0.1 − 0.2	2	0.25 − 0.5	2
0.2 − 0.3	3	0.5 − 1	5
0.3 − 0.4	4	1 − 2	9
0.4 − 0.5	5	2 − 3	12
0.5 − 0.6	6	3 − 4	14
0.6 − 0.7	7	4 − more	16
0.7 − 0.8	8		
0.8 − 0.9	9		
0.9 − 1	10		
1 − 2	11		
2 − more	12.5		

extremely good concessions the difference between the two versions becomes larger, up to about 4%. Such situations, however, are unlikely in the North Sea area. The difference between the original and the approved decree is consequently only minimal. The average royalty to be paid in the "liberal" decree is slightly higher for large outputs than in the "socialistic" version. The same is true of the natural-gas regulations.

Taxation

The taxation provisions in the Decree of January 1967 include an investment-credit of 10%. The legislation of the original draft also contained a credit, but it was 10% of the gross profit. The latter provisions are clearly more attractive for the companies if large and prolific fields can be found; the former regulation is better if marginal fields are discovered. Since the majority of the fields are generally marginal or providing a small profit, the investment credit is better suited to stimulate exploration than the credit accounted as 10% of the gross profit. Which of the two regulations is the best for the companies depends entirely upon the geological expectation, and the geological model that is the basis for the exploration strategy. It must be mentioned that the present value of the investment-credit is larger than the present value of a comparable amount of money earned as 10% of the gross profit. The investment-credit can be earned directly at the beginning of the project, before the pay-out time is reached. The 10% of the gross profit, however, can be subtracted only after the pay-out time is fulfilled.

State participation

The major differences between the two versions were on the subject of the state participation. The "liberal" decree provides only for state participation in the case of dry-natural gas, of 40%. The "socialistic" version included state participation for both oil and gas. This form of state participation was introduced with a number of accessory provisions that facilitated the companies' acceptance of the legislation; that is, in the case of oil and of gas, a "handling fee" and a royalty-credit for the companies was included in the provisions.

The regulation for natural gas was that the state could participate after the discovery of an economically-recoverable reserve of natural gas. The amount of participation could never exceed 50%. If the state participation was established at $a\%$, a royalty-credit was granted in which the original sliding scale had to be diminished by the fraction $\frac{100 - a}{100}$. For instance, if a state participation of 25% was realized, only 75% of the original amount of royalties had to be paid. The effect of the royalty-credit is larger for concessions with a extensive production than for concessions with a limited one. Assume for instance a production of 8×10^9 m^3/year. Without state participation the royalty to be paid would have been 12.75%, and with a state participation of 50% only 6.875%. In this case the licensee is producing 4×10^9 m^3/year on its own account and the other half for the benefit of the state. If the licensee, however, should have produced 4×10^9 m^3/year from a concession without state participation, the weighted average of the royalty would have been 9.5%. The royalty-credit is consequently exactly 2.625%. This does not apply (even the reverse is true) to small outputs. A production of 0.5×10^9 m^3/year inclusive state participation of 50% would require an average royalty of 0.5%. Without the state participation, however, a production up to 0.25×10^9 m^3 per year is free of royalty! The royalty-credit for natural gas is higher in the accepted Decree of 27 January 1967. In this decree the state participation was restricted to 40%, while 50% of the royalty could be subtracted as a royalty-credit. So for the higher outputs, the royalty-credit is substantially larger in the accepted decree. In the "socialistic" version the royalty-credit is proportional to the degree of state paticipation; and a state participation of 40% would have resulted in a royalty-credit of 40%.

On the other hand the "socialistic" version included a handling fee for the licensee who was obliged to handle the natural gas produced for the state's account. This handling fee was established according to a sliding scale, which is given in Table XVIII. Above a production of 5×10^9 m^3 per year, or about 175×10^9 cuft. per year, no handling fee was granted. This handling fee is in fact a negative royalty—calculated on the

TABLE XVIII

SLIDING SCALE FOR THE HANDLING FEE OFFERED TO THE COMPANIES FOR THE TREATMENT OF NATURAL GAS WHEN STATE PARTICIPATION IS EFFECTUATED, AS PRESENT IN THE UNACCEPTED VERSION OF THE GOVERNMENT-CALS FOR THE DUTCH OFFSHORE-PETROLEUM LAW

production/year	perc. of sales value
less than 1×10^9 m^3	5
$1 - 2 \times 10^9$ m^3	4
$2 - 3 \times 10^9$ m^3	3
$3 - 4 \times 10^9$ m^3	2
$4 - 5 \times 10^9$ m^3	1

basis of the sales price—and paid by the state to the licensee, to cover production that can be regarded as a property of the state. In the case of production up to 1×10^9 m^3 or 35×10^9 cuft. per year, and with a state participation of 50%, this meant that the state shared for 50% in the costs of the limited-liability company, but for only 47.5% in the gross revenue. The state participation for gas is in the "liberal" decree characterized by a 40% participation and an important royalty-credit for the larger outputs, and in the "socialistic" version by a participation up to 50% and a somewhat lower royalty-credit, but which includes an important "handling fee" for the smaller outputs. The differences between the two versions with regard to the state participation in natural-gas projects are consequently slight. It depends partly on the output which of the two the companies prefer.

The possibility of state participation in oil-production projects existed only in the "socialistic" version.

For oil as well the state could participate for a percentage up to 50%. The royalty-credit was, however, very important because the royalty to be paid was established according to Table XVII—but the fraction that could be applied to calculate the royalty-credit was $\frac{50 - a}{50}$, implying that for a state participation of 50% no royalty was required, whatever the output!

Apart from the royalty-credit an important handling fee was included. This handling fee amounted to 15% of the sales price of the output. This meant that if the state participated for 50% in the costs, its participation in the gross revenue was only 42.5%. The financial burden of the state participation was eased by a high royalty-credit and by the handling fee.

The handling fee and the royalty-credit were so important in this version of the decree that it can be doubted whether state participation could have been profitable in most cases. An example is when the ratio between revenues and costs is 100 : 50, the royalty 9%, and the corporate income tax 45%. The proceeds for the state would be 27.5 units without state participation and 32 units with a state participation of 50%. This would have meant that the *incremental* benefit from the state participation would be 4.5 units, for which the state had to invest 25 units. The company is earning 18 units in the case of state participation for the same amount of investment. In this instance it can be concluded that the profitability of the investment for the state is 25% of that of the company. Clearly this incremental benefit to the state would in most cases be insufficient to justify state participation. Only in the case of exceptionally rich fields—for instance, with a revenue: costs ratio of 100 : 20 is the profitability of the investment for the state roughly half (depending on the royalty) that of the company. Since the rate-of-return on exceptionally rich fields will be high, however, it can be assumed that in such cases state participation would be profitable to the state. Since the chance of finding bonanza's is extremely small, this type of state participation could have influenced only the very end of the "tail" of the expectation curve. The state participation for oil could not have had under these additional terms any major influence on the expected exploration activity, since the companies should have realized the fact that the state would participate only in case of the discovery of a bonanza. Consequently the state participation for oil could not have had an important influence on the companies' attitudes—at least when considered purely as an appraisal of project investments.

Comparison of the two versions of the decree

From the standpoint of the companies, the two versions appear very much the same. Bonus and surface duties are identical in both. The scale for the royalties is for oil and gas slightly more favorable in the "socialistic" version than in the "liberal" decree. The difference in taxation depends on the geological expectation, and this difference is also minimal. The major differences are found in the state participation. For natural gas the "socialistic" version included a participation of 50%, an important royalty-credit and a handling fee. The "liberal" decree placed the state participation at 40%, included a more generous royalty-credit, but excluded the handling fee. Only the "socialistic" version included a possibility of state participation for oil, yet the terms were such that state participation could only conceivably be realized when a bonanza had been discovered. We can thus conclude that the differences between the two decrees are only slight.

Since both versions are based on the same act, the Continental-Shelf-Mining Act, the "socialistic" and the "liberal" varieties show few real differences. This conslusion is somewhat ironic, considering the high level of activity spent by the Dutch Parliament and Government in changing the "socialistic" version into a "liberal" one. The final situation implies that is was a rather wasted Parliamentary effort. The extensive leasing activity of the companies which developed as soon as the decree was accepted proved that the geological expectation of the companies was such that the oil legislation was completely acceptable to them. Since the "socialistic" version was only marginally different, the companies would also have accepted this legislation. But in this case the exploration activity would have begun earlier, which would have brought accompanying increased profits for the Dutch State.

Conclusions

From the previous sections it can be concluded that the Dutch oil legislation for the continental shelf is in close accordance with the criteria that were established in the previous chapter (VI). This is true noth of the approved legislation and of its predecessor, the "socialistic" version. Both versions lead undoubtedly toward a level of earnings for the state that approaches the maximum public revenue. The differences between the two versions are from a financial standpoint minimal.

Literature

References

Jonkers, J., 1966. *Mijnwetten—Nederlandse staatswetten.* Tjeenk Willink, Zwolle, 281 pp.

Statute Book of The Kingdom of The Netherlands, 1965. Act of 23 September 1965 embodying regulations with regard to the exploration for and the production of minerals in or on the part of the Continental Shelf that is situated underneath the North Sea (Continental Shelf, Mining Act). *Staatsblad 428.*

Statute Book of The Kingdom of The Netherlands, 1967. Decree of 27 January 1967, pursuant to Article 12 of the Continental Shelf Mining Act as regards exploration and production licences, inter alia for petroleum or natural gas. *Staatsblad 24.*

Selected documents

Bouchez, L. J., 1966. Economische aspecten van de mijnwetgeving voor het continentaal plat. *Econ. Statist. Ber.*, pp. 910-912.

Driessen, H. E. A., 1964-1965. Het ontwerp mijnwet continentaal plat. I t/m VI. *Econ. Statist. Ber.*: 14 okt. 1964, 21 okt. 1964, 28 april 1965, 12 mei 1965, 26 mei 1965, 9 juni 1965.

Driessen, C. F., 1966. De mijnwetgeving op de Noordzee. Tijdschrift v. h. Kon. Ned. Geol. en Mijnb. Genoot., *Geol. Mijnbouw,* 45: 75–82.

Schierbeek, P., 1965. *Olie en Gas in Nederland en onder de Noordzee.* Amro-Bank, 38 pp.

Van Meurs, A. P. H., 1966. Hoe verder met het Nederlandse "natte" mijnbouwbeleid? Deel I: *Econ. Statist. Ber.*, 2564: 1079-1082. Deel II: *Econ. Statist. Ber.*, 2565: 1108-1113.

Chapter VIII

The offshore-mining legislation in other industrialized countries

A short study of the offshore-mining legislation of some other industrialized countries may yield valuable observations which can be used to check the results of our study in the Dutch offshore-mining legislation. Such a study is necessarily incomplete. Each state has its own peculiar problems and consequently its own mining policy. The geological expectations upon which government bases its mining legislation vary, as well as the economic and technical situation. A complete and thorough analysis of every mining legislation is therefore a study so extensive as to be beyond the scope of this thesis. This does not indicate, however, that a limited comparison is of no value.

This study will develop in two steps. First the offshore-mining legislation of other North Sea countries will be studied and compared with the Dutch legislation. Secondly, petroleum legislation in other industrialized states, such as the U.S.A. and Canada, will be analyzed.

Offshore-mining legislation in the North Sea countries

Economic conditions comparable to those of The Netherlands prevail on the other North Sea countries, lending added value to a comparison with them. Four states will be studied: Great Britain, Norway, Denmark, and Western Germany. In Belgium, no offshore-mining legislation exists.

The offshore-mining legislation of the four other countries originated before 1965, or before significant exploration was underway in the North Sea area. The Dutch legislation, however, was approved three years later, August 1967. The Dutch legislation was constructed with more geological information concerning the North Sea, after the first results of extensive exploration on the British side of the shelf were known.

The offshore-mining legislation of Great Britain, Norway, Denmark, and Germany was based on considerably less information, consisting almost exclusively of some indications from the onshore geology and geophysical indications offshore. Some further technical and economic facts that were known were the distance from important markets, the average water depth in the North Sea, and the possibilities for pipeline construction to shore.

The distance from the important markets for oil and gas was in Great Britain and Western Germany comparable to that of The Netherlands. In these three countries important markets for oil are available. The Netherlands' gas market is entirely supplied by the Slochteren gas field and any newly-discovered gas must be exported; but in Belgium, Western Germany, and France important additional quantities can be supplied. The gas markets of Great Britain and Western Germany are open to large new quantities of gas. In Denmark and Norway potential consumers are considerably fewer, while the gas must be piped for considerable distances before these relatively small markets can be reached.

The average water depth in the North Sea is rather shallow in the Dutch, German, Danish, and most of the British shelf. Large areas do not exceed 50 meters (165 ft.). Between the centre of the North Sea and the Norwegian coast is situated a deep trough, as can be seen on Fig. 40.

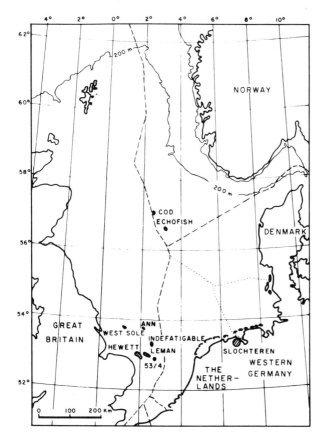

Fig. 40. Discoveries in the North Sea area

This trough engenders great difficulties for an eventual pipeline construction to the Norwegian coast, making transport conditions unfavorable for natural gas. Transportation of natural gas does not display comparable drawbacks in other parts of the North Sea.

SCHIERBEEK (1965) gives an excellent review of the petroleum legislation in the North Sea countries and the following analysis will be based on his article.

Great Britain

The British petroleum legislation is based on the Continental-Shelf Act 1964 and the regalian system. The state does not own the minerals but has the right to regulate exploration and production. It is possible to obtain a non-exclusive exploration licence and an exclusive production licence.

The non-exclusive exploration licence includes the possibility of drilling wells to a depth of 1,000 ft. (330 meters).

The surface of the blocs in the exclusive production licences is 100 sq. miles (250 km^2). Duration of the licence is 40 years. After 6 years, 50% of the area must be relinquished. The work obligation is negotiated before the licence is granted.

The financial aspects of this legislation include the payment of a fixed-initial bonus of £ 200.— for each of the first ten blocs and £ 5.— for each subsequent bloc.

The surface duties amount to £ 25.— per km^2 for the first six years, and £ 40.— in the seventh year; each following year the surface duty increases by £ 25.— to a maximum of £ 290.— per km^2 after seventeen years.

The royalty is 12.5% of the wellhead price based on the sales price. The royalty is consequently fixed.

The corporate-income tax is 40% of the taxable income. An investment-credit of 30% exists for fixed assets and intangible drilling expenditures.

No state participation is required.

This set of financial conditions falls in line with a situation in which the government expects a relatively low probability-of-success ($0.01 < p < 0.1$). The conditions pressure will be on the average as proposed in conclusion 4 (c) in Chapter VI. The fixed royalty of 12.5% will probably eliminate a few marginal prospects; but since the bonuses and the surface duties are low, it cannot be expected that a large number of projects were dropped by the companies due to the severe mining legislation. The tax regulations are somewhat favorable to the companies, since the tax rate is low and the investment credit is attractive.

These conditions lead us to the conclusion that a considerable rent will accrue to the companies when exploration proves successful.

September 1965 saw the first successful exploration well which further investigations proved to lie in the West Sole field. This altered the geological picture completely. A Labour government had in the meantime succeeded the Conservative government; a declaration of the former was issued which declared that those companies that were ready to accept a joint venture with a British state company for exploration and production would be favorably treated in the application for licences for new blocs. This resulted in the joint venture between AMOCO and GAS COUNCIL for the blocs awarded in November 1965. The other conditions, however, remained unchanged.

An important variable, however, remained to be decided upon and was not specified in the contracts: the price to be fixed for natural gas. All the gas had to be offered to the GAS COUNCIL, and this state-owned company consequently had a monopolistic position. The first contract was between the GAS COUNCIL and the BP for the gas supply from the West Sole field. The price was 5 d/therm. This price was relatively high, yet at about the same time (summer 1967) the gas from Algeria transported to Canvey Island terminal in London was cut off by the upheavals in the Arab-Israëli war. British gas reached the coast exactly in time to cope with these problems, explaining the high gas price.

The second gas contract was between the GAS COUNCIL and the PHILLIPS-group for the supply from the Hewett field. The price was agreed upon at 2.87 d/therm. This price, however, was established after the devaluation of the British pound, and was low related to the price of the Slochteren gas. Similar gas contracts were established with SHELL/ESSO and the AMOCO-group.

It can be concluded from these observations that a considerable part of the rent is presently earned by the British state in enforcing this relatively low gas price.

The history of the British offshore mining legislation can be described as follows: mining legislation was established that was first rather profitable for the companies. This resulted in considerable exploration activity, which proved to be successful. At that moment two mediums, the option for a joint venture with the state and the gas price, resulted in the fact that a large portion of the rent was earned after all by the British state. The British government proved to be a master in fishing behind the nets!

Since in 1970 large parts of the original blocs must be returned to the state, it is likely that the government will study the possibility of changing the petroleum legislation. Political pressure has recently built up demanding a larger role for state-owned companies.

Norway

As in Great Britain, there exist in Norway two types of licences: non-exclusive exploration licences and exclusive production licences. The non-exclusive exploration licences do not include the possibility for drilling.

The surface of the blocs with the exclusive production licences is about 215 sq. miles (550 km^2). The duration of the licence is 6 years, but it can be extended to 40 years. After 6 years, 25% of the area must be relinquished and after 9 years another 25% of the original area. The work obligation must be negotiated before the licence is granted.

The financial arrangements include a fixed initial bonus of Kr. 10,000.— (about U.S.$ 1,600.— before devaluation).

Surface duties may be deducted from royalty payments if exploration results in production. Surface duties are Kr. 500.— per km^2 for the first 6 years. Afterwards the duties increase to Kr. 5,000.—/km^2 with yearly jumps of Kr. 500.—.

Royalties amount to 10% of the wellhead price. Corporate income tax amounts to 42.25% of the taxable income. No special tax credits are granted.

The financial conditions of the Norwegian petroleum legislation agree with a conditions pressure proposed in 4(c) in Chapter VI (generally-expected probability-of-success between 0.01 and 0.1). The conditions are less favorable to the companies than in the British legislation, but the difference is only slight. Since, apart from the mining legislation, the technical conditions on the Norwegian shelf are also less favorable (mainly due to the natural-gas transport problems), the conclusion is valid that the conditions for the companies are less appealing on the Norwegian side of the North Sea than on the British side.

Until now the exploratory activity has not proved particularly successful. Only the PHILLIPS-group met with success with the Cod field, a natural-gas with condensate area;

however, these resources are situated almost beneath the middle of the North Sea. Recently rumours were spread that the gas would be sold to the U.S.A. and be transported with methane tankers. The same group has found an oil field, the Echofish oil field, in about the same area.

Denmark

Exploration and production of the continental shelf of Denmark has been awarded to a consortium, the "Dansk Undergrunds Consortium". In this consortium participate a Danish shipping magnate A. P. Möller, and GULF, SHELL, TEXACO, and California Oil Company.

The concession embraces the entire Danish part of the shelf, and its duration is 50 years. There are no bonuses, surface duties or state participation. The royalty is 5% for the first 5 years and 8.5% for subsequent years, calculated on the wellhead value.

The corporate-income tax is 44%, but half of the tax paid in the previous year may be subtracted for the calculation of the taxable income. This makes the ultimate level of the income tax about 36%.

The financial conditions in the petroleum legislation in Denmark are in accordance with a generally expected probability-of-success between 0.01 and 0.1 (as explained under 4(c) in Chapter VI). The conditions on the Danish side of the continental shelf are the most favorable of any of those imposed upon companies by countries in the North Sea area. This low conditions pressure corresponds to the low geological expectations for this portion of the North Sea continental shelf.

No important discoveries have been made on the Danish side of the North Sea as of now.

Western Germany

The right to explore and produce from the Western German continental shelf was granted originally to a consortium, the "Nordsee Konsortium", in which AMOCO, BRIGITTA, MOBIL, and a large number of German and French companies participated. The consortium relinquished a number of blocs that were subsequently granted to companies in 1965 with the right to explore until October 31, 1969.

The financial arrangements, as far as is known, do not include bonuses or surface duties. Royalties are expected to be 5%. The Konsortium has negotiated a work obligation. No state participation exists. Corporate-income tax is about 45%. Western Germany is stimulating oil and gas exploration outside its territory (the continental-shelf area is regarded as being situated outside German territory) by interest-free loans that must be repaid only in the case of a commercial discovery.

The financial aspects of the German petroleum legislation fall clearly in the range in Chapter VI described as belonging to the category of a generally expected low probability-of-success. Although some exploration was performed on the German platform, the results were disappointing. No commercial producers have as yet been found. The activity is also diminished, however, by the juridical struggle concerning the demarcation of the platform between Western Germany and Denmark, and Western Germany and The Netherlands.

Comparison of the offshore petroleum legislation of the various North Sea states

A striking difference exists between the Dutch legislation on the one hand and the legislations of the other North Sea countries on the other.

Exploration was until now almost a complete failure on the Norwegian, Danish and German continental shelf. Only the British shelf proved fruitful. This produced in 1967 a rather optimistic expectation for the Dutch offshore—situated between the largest onshore gas field, Slochteren, and the largest offshore gas field, the Leman gas field. The geological expectation which then existed can be best illustrated with a quotation from WORLD OIL (July 1968, p. 88): "Ultimately, the North Sea may yield up to 300 trillion cubic feet of gas and some oil", and on the same page is written: "Explorationists will soon know whether excellent Dutch offshore geophysical prospects are major gas fields as several are in the UK part of the North Sea—or whether the Rotliegendes Sandstone is eroded off the tops of the structures or changes to salt as it does offshore from West Germany."

This optimistic view resulted in petroleum legislation with harsher conditions for the companies on the Dutch offshore. It is understandable that the companies accepted the Dutch terms, since the geological expectation was better in 1967 for the Dutch offshore than in 1965 for the other areas, justifying the Dutch policy. It must be mentioned, however, that the firmer Dutch terms only apply to the bonanza's and the intermediate fields. The Dutch terms are more favorable for marginal fields than in other countries bordering the North Sea. Fig. 41 illustrates the conditions pressure as a function of the ratio between costs and revenue. The low conditions pressures for marginal projects in the Dutch legislation are clearly evident. The low conditions pressures originate from the sliding-scale royalties, in combination with the tax credit of 10%.

Note that the graphs in Fig. 41 show only a general tendency, because the time pattern of earnings by the government and the companies, and consequently the present value of the earnings, will differ for each project, as will be clear from the analysis in Chapter VI.

Another attraction of the conditions on the Dutch side of the continental shelf was the expected price for natural gas. Since most of the eventually-discovered gas needs to be exported, the Dutch government can be expected to assist in anything that may prevent the gas prices from eroding. Gas prices, therefore, can be expected to be comfortably high.

The differences among legislations from the other four countries are slight. Since the geological expectation is low for the Danish shelf, terms highly favorable to the companies are justified. There is a discrepancy between the Norwegian terms and the British terms, because the former are slightly more discouraging while the geological, technical and economic conditions on the British shelf are better. Since oil prices can be expected to be about the same in Great Britain and Norway, the present British terms are clearly too low for oil in relation to those from Norway. Very little is known about the

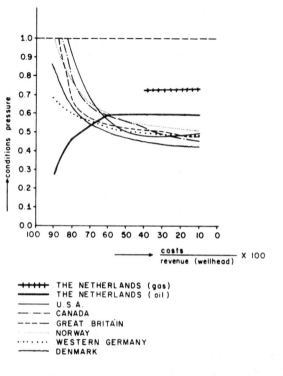

THE NETHERLANDS (gas)
THE NETHERLANDS (oil)
U.S.A.
CANADA
GREAT BRITAIN
NORWAY
WESTERN GERMANY
DENMARK

Fig. 41. Conditions pressures of mining legislation from various industrialized countries

expected Norwegian natural-gas prices, but even if those prices are slightly higher, it must be assumed that the British terms are still too low for natural gas related to the Norwegian ones.

The difference in geological attractiveness between the British and the Norwegian shelf may lessen if the Echofish (Ekofisk)-oil field proves to be a giant and prolific one. Exploratory activity may even be drawn towards the centre of the North Sea area, due to a new favorable geological expectation.

Technical conditions for oil are comparable for both shelves because oil can be transported by tanker instead of with a pipeline, avoiding the deep trench between the shelf and the Norwegian coast.

Consequently, if the development of the Ekofisk-oil field proves successful the difference between the Norwegian and British conditions is only slight, and the previous conclusion is less valid.

Exploratory results on the Danish and German shelf are very disappointing. No commercial gas or oil fields have been discovered. Since the conditions pressure will be on the average low for these areas, little additional activity can be expected from a change in the mining legislation in these countries. The exploratory activity may revive if the juridical struggle between Denmark and Germany is terminated and important discoveries are made in the areas that were disputed until the present.

Exploratory results on the Dutch side have also been somewhat disappointing. Of the 24 exploratory wells only one has found gas in sufficient commercial quantities to warrant the application for a production-licence, although three other wells have also found gas in significant quantities.

On the other hand, considerable interest existed early in 1969 for the remaining blocs (ANONYMOUS, 1969b), indicating that companies still regarded the Dutch shelf as sufficiently profitable for the given terms at that time.

In the Dutch offshore as well a promising oil discovery was made by TENNECO (bloc F 18). Therefore, a renewed interest may arise for exploration in the northern part of the Dutch shelf if the territorial dispute with Western Germany is settled.

For the moment, there is no reason to change the terms for new applications.

The U.S.A.

The mineral legislation in the U.S.A. dates from 1866, when the first essentials were incorporated into a law. In 1872 the Mineral-Location Law became the cornerstone of the American mining legislation (FLAWN, 1966, p. 167). A central principle is that the landowner of the surface has the right to earn proceeds from the mining operations. In petroleum exploration and production, landowners earn bonuses, rentals, royalties or aquire other rights as for instance a carried interest, by transferring the minerals' owner-ship to a petroleum company.

The Federal Government is only earning taxes from the activity on private proper-ties. For the areas of Federal influence, such as the continental shelf, the Federal Govern-ment acts as a landowner by requiring the various financial compensations.

Since it can be safely assumed that most landowners try to maximize possible revenues from petroleum operations, a type of "free-market mechanism" exist in the U.S.A. for the establishment of lease conditions. Landowners earn high bonuses, rentals, or royalties in areas regarded as favorable by the petroleum companies, while low amounts are paid in areas of marginal interest. Landowners earn in most cases a signifi-cant part of the expected true rent from the petroleum production. The bonuses paid vary considerably among the landowners; rentals are one or more dollars per acre while royalties range from 10 to 20% in most cases.

Apart from these payments to the landlords, taxes are due to the Federal and State governments.

The Federal-corporate-income tax

The Federal-tax rate in the U.S.A. for corporations is about 50% of the taxable income. The main source of income for most petroleum-producing companies is, natural-ly enough, through the sale of petroleum. The gross income in the U.S.A. is calculated on the basis of the price for which the taxpayer sells the oil and gas in the immediate vicinity of the well. The taxpayer is accounted only for his own interest in a producing property. The royalty, for instance, can be deducted in calculating the gross income (MILLER, 1963).

From the gross income the operating expenses and the yearly capital costs can be deducted to arrive at the taxable income. The operating expenses are outlays for rentals, salaries, fuel, repairs and maintenance, etc. Capital costs can be divided into three categories:

(a)   those that represent investments in objects that have no salvage value—these capital investments can be deducted immediately in the year of the outlays for tax purposes;

(b)   those that represent investments in objects that have a salvage value—these costs must be depreciated for tax purposes;

(c)   those that represent investments in capital that is consumed during production— these costs must be recovered through depletion for tax purposes.

In the tax legislation in the U.S.A., outlays for exploratory drilling resulting in dry holes, geological and geophysical costs that do not result in productive properties, all exploratory-drilling costs, and intangible development costs can be written off immediately. Tangible development costs must be depreciated. Lease bonuses and geological and geophysical costs attributable to a productive lease must be recovered through depletion.

The companies can choose to use a cost depletion or a percentage depletion. The percentage depletion is 22% of the gross income or 50% of the taxable income before deduction of the depletion allowance, whichever is less. The percentage depletion gradually developed from the costs depletion and was introduced mainly for its administrative simplicity as compared to the cost depletion (LENTZ, 1960). It must be mentioned that 22% of the gross income cannot ultimately be earned by a company, since at the beginning and the end of the lifetime of a well the differences between the gross income and the net income are too small to allow the 22% deduction, or the well is operating at a loss and consequently no deduction is possible. Therefore, the actual deduction is roughly 20% of the gross value of the production. Clearly, the depletion allowance of 22% yields far greater benefits than the bonus and geophysical and geological costs attributable to the lease; therefore, most companies choose the percentage depletion instead of the cost depletion.

The possibility for direct deduction of exploratory-drilling costs, dry-hole costs, and bonuses paid for unproductive leases, in addition to the provisions for this depletion allowance, can together be regarded as highly favorable tax treatment as compared to other industries. In fact, it consists of the grant of a subsidy to the petroleum industry. This special tax treatment is surely one of the reasons that the petroleum industry in the U.S.A. has developed so enormously, and correspondingly, why the U.S. companies have gained such enormous influence in the international petroleum market (ARPS, personal communication). The appeal of the tax treatment in the U.S.A. is clearly illustrated by the recent merger of SOHIO and BP, a deal that allowes BP to offset Alaskan drilling costs against the profit-making marketing of SOHIO in the U.S.A., for tax purposes (ANONYMOUS, 1969).

Originally, the depletion allowance amounted to 27.5% of the gross income, as established by the Revenue Act of 1926. Political pressure, mainly from circles not engaged in petroleum exploration and production, resulted in a recent change into a 22% depletion, along with a number of other changes in the tax law that negatively influenced the tax position of the petroleum companies. KINNEY (1969) accounts that the loss to the industry has been about U.S.$ 600.— x 10^6. However, it must be admitted that the 22% depletion is still a powerful subsidy.

Conditions for Federal leases or State leases

Three different financial arrangements dominate the conditions for most Federal or State leases: the bonus, royalties, and the income tax. The last element has been discussed in the previous paragraph; the bonuses and the royalties remain to be handled.

Bonuses. In most cases bonuses are established by bidding. For the Federal-lease sales of offshore Texas and Louisiana, and for the State-lease sale in Alaska, this system was applied. The average price paid by the successful bidders in Alaska was U.S.$ 2,000.— per acre or U.S.$ 1,280,000.— per sq. mile! These prices are enormous as compared, for instance, with the fixed bonuses paid on the Dutch offshore of about U.S.$ 1.— per acre. This price was, however, an average one because the highest bid was U.S.$ 28,244.— per acre for the area adjacent to the lease with the famous Prudhoe Bay discovery well. The payment of a bonus equal to 14 times the average price for "good" blocs illustrates clearly the importance of the geological expectation in the conditions pressure that a company is willing to accept. The bids in Alaska were extraordinary high. The offshore sale in Texas resulted in bonuses at an average of U.S.$ 1,100.— per acre. This price, however, is high with regard to prices paid in the past, because in similar sales offshore Texas in 1960 only U.S.$ 150.— per acre was paid (SCOTT, 1969).

BARROW (1966) mentions that outlays for lease bonuses constitute a very important part of exploratory investments. For instance, the total exploratory investments on offshore Louisiana amounted between 1951 and 1965 to U.S.$ 1,600.— x 10^6, of which U.S.$ 1,055.— x 10^6 was paid for bonuses! Barrow also points out that these high bonuses involve the companies in considerable risks. For instance, one company spent U.S.$ 50,— x 10^6 offshore Louisiana on bonuses, but in the previously mentioned period, found no area meriting commercial production.

It can be argued that these enormous bonuses distrturb orderly competition between petroleum companies. These large sums can be paid only by large companies, with the result that the smaller companies are sold out by the larger ones. But even for large companies the payment of these bonuses constitutes a large drain on the possibilities for investment, without any substantial result in exploration. Since in new environments, where sufficient geological information is lacking, such as the Alaskan North Slope and the offshore Texas Louisiana, the discovery of a petroleum accumulation is to a large extent due to luck, it seems that the financial position of a petroleum company in the U.S.A. is, to a large extent, determined by luck. This does not meet the requirements for orderly competition.

Royalties. The royalties for Federal leases are $16^2/_3$% of the value of the production. These are rather heavy requirements. No sliding scales exist for Federal leases in the U.S.A.

Proration is also important for the evaluation of the future profitability of petroleum prospects. In the Gulf Coast area, these proration measures are of great importance, as is illustrated in the article of ARPS (1961).

Canada

The provinces in Canada have mining laws that differ widely. The provinces have the right to sell public lands, of which they hold a relatively large percentage. Leases offered by the provinces are called "Crown leases". Most provinces raise taxes apart from the Federal taxes. Provinces regulate the proration.

Federal taxation

The Canadian Federal-corporate-income tax differs in many aspects from that in the U.S.A. All exploration expenses, whether resulting in a productive lease or not, can be deducted in the year of expenditure. Expenditures outside Canada, however, cannot be deducted—except those that can be deducted from income rising from production of a well outside Canada. In the U.S.A. all expenditures outside that country are equally deductable as those within it.

All mining companies in Canada are in the first three years exempted from Federal income tax. The tax rate is, as in the U.S.A., 50%. Contrary to the U.S.A., the depletion allowance is $33^1/_3\%$ of the taxable income before depletion allowance (HODGSON and BEARD, 1966). For income from royalties, the depletion allowance is only 25% of the net income.

Crown leases

In Canada as well, bidding is used to fix the bonuses for the Crown leases. Royalties are in most provinces according to a sliding scale (MATTHEWS, 1963). An example is Alberta, where the royalty is 8% up to 25 barrels per day; the royalty to be paid over the following increasing production is higher and reaches $16^2/_3\%$ for a production of 100 barrels per day and more. On freehold leases a fixed royalty of 12.5% is normally paid.

Comparison with the U.S.A.

If taxes paid to the provinces are omitted from the analysis, the Canadian financial treatment is more favorable for the marginal fields due to the sliding-scale royalties. For the intermediate fields the financial arrangements in the U.S.A. are more advantageous because of the possibility of subtracting at least 50% of the net income as a depletion allowance. For the bonanza's, the Canadian treatment is again more favorable since the limitation of 22% of the gross income does not exist in Canada.

Therefore, on the average the practices in the U.S.A. and Canada will result in about the same conditions pressure for comparable cases, as is illustrated in Fig. 41.

Comparison of the petroleum legislation in industrial countries

If bonuses are omitted, the conditions pressure for most of the projects in industrialized countries will be remarkably similar, as can be seen in Fig. 41. As long as the costs are beneath 60% of the wellhead value of the oil, the conditions pressure ranges from about 0.4 to 0.6. The conditions pressure is lowest in Denmark and highest in The Netherlands. Since very large sums are spent in the U.S.A. and Canada for bonuses, it can be safely assumed that the conditions pressure for the projects in this range is considerably higher in North America. For instance, if the bonus amounts to 5% or more of the wellhead value of the oil, the conditions pressure for the bonanza's is approximately equal to that in The Netherlands, and is substantially higher for the intermediate fields. The important aspect of these bonuses, however, is that they are offered by the companies themselves and the conditions pressure will therefore never exceed 1 throughout the bidding for these bonuses.

For natural gas the conditions pressure is considerably higher in The Netherlands than in other industrialized countries where state participation is applied.

As mentioned previously, the conditions pressure for marginal projects in The Netherlands is very low compared to other petroleum laws.

From this it can be concluded that the Dutch mining law is undoubtedly better suited to realizing the maximum public revenue than other national legislations. A large portion of the rent is earned from the profitable projects, while at the same time the economically-recoverable reserve will be larger in The Netherlands than elsewhere, due to the favorable treatment of marginal projects.

If, apart from the financial aspects of mining legislation, economic and technical conditions and market regulations are considered, a comparison between the U.S.A. (e.g., the Federal-offshore leases in the Gulf Coast) and the North Sea countries becomes more difficult. The technical possibilities for oil and gas production are similair in both offshores although the weather conditions are not favorable on the North Sea. The price for oil is high in the U.S.A. and low in Europe; for natural gas the reverse is true. The distance to important consumer centres is rather large from the Gulf Coast area, while the North Sea is in the middle of a quick-growing market for oil and gas.

Proration measures may have a highly negative influence on the profitability of exploratory ventures. Proration does not exist in Europe.

From this list of differences it can be safely concluded that a proper comparison between the conditions in the U.S.A. and western Europe is very difficult.

In both regions, legislation is rather selective, which will result in the production of as many fields as possible. The conditions pressure in the U.S.A. is heavily determined by the bonus. On the average the conditions for oil seem better in the Gulf Coast area, mainly due to the high market price for oil, while the conditions for natural gas seem on the average better in western Europe.

Literature

References

Anonymous, 1968a. North Sea gas price agreement causes furore. *World Petrol.*, 39(5): 20H-20I.

Anonymous, 1968b. Commercial finds elude Norwegians. *World Petrol.*, 39(11): 27-28.

Anonymous, 1969a. Norway claims Phillips has a big one. *Oil and Gas J.*, 67(45): 130-131.

Anonymous, 1969b. Dutch awards end NAM's monopoly. *Offshore*, 9(4): 93-94.

Anonymous, 1969c. Why the bids were high. *Petrol. Press Serv.*, 36(10): 362-363.

Arps, J. J.,1961. The Profitability of Exploratory Ventures. In: INTERNATIONAL OIL AND GAS EDUCATIONAL CENTRE SOUTH WESTERN LEGAL FOUNDATION (Editor), *Economics of Petroleum Exploration, Development, and Property Evaluation*. Prentice Hall, Englewood Cliffs, N. J., pp. 153-173.

Barbeau, J., 1963. The rationale of the Canadian treatment. *Oil and Gas Production and Taxation*: 256-277.

Barrow, T. D., 1966. Economics of offshore development. In: INTERNATIONAL OIL AND GAS EDUCATIONAL CENTRE SOUTH WESTERN LEGAL FOUNDATION (Editor), *Economics of the Petroleum Industry; New Ideas, New Methods, New Developments*. Gulf Publ. Comp., Houston, pp. 133-146.

Flawn, P. T., 1966. *Mineral Resources. Geology—Economy—Engineering—Politics—Law*. Rand, Mc. Nally and Company, Chicago, New York, San Francisco, 406 pp.

Hodgson, E. C., and Beard, W. J., 1966. Summary review federal taxation and legislation affecting the Canadian mineral industry. *Dept. of Mines and Techn. Surveys, Ottawa. Mineral Inform. Bull.*, 82 MR: 1-25.

Kinney, G. T., 1969. Conferees, tax bill pumps industry for $ 600 million. *Oil Gas J.*. 67(52): 73-75.

Lentz, O. H., 1964. Mineral economics and the problem of equitable taxation. *Quart. Colo. School Mines*, 55(2): 1-111.

Matthews, M. W., 1963. Operating methods in the oil and gas industry. *Oil and Gas Production and Taxes*: 92-135.

Miller, K. G., 1963. The United States income taxation of the industry. *Canadian Tax Papers*, 33: 200-238.

Schierbeek, P., 1965. *Olie en Gas in Nederland en onder de Noordzee*. AMRO-Bank, The Netherlands, 38 pp.

Scott, J., 1969. Texas offshore: breakthrough in the making. *Petrol. Eng. Intern.* : 53-56.

Wells, M. J., 1968a. UK North Sea under financial pressures. *World Petrol.*, 39(3): 40-47.

Wells, M. J., 1968b. State oil entity, slow drilling, mark UK autumn. *World Petrol.*, 39(12): 30-34.

Selected documents

Ely, N, 1961. Summary of Mining and Petroleum Laws of the World. *U.S. Bureau of Mines Information Circular* 8-17, 215 pp.

Ely, N., 1964. Mineral Titles and Concessions. In: *Economics of the mineral industries*. (2nd ed.) American Institute of Mining and Metallurgical Engineers., New York, pp 81-130.

Petrol Taxation Report, 1965 etc. A Monthly Digest and Analysis of Fiscal Developments Affecting Petroleum Payments around the World. New York.

Sullivan, R. E., 1955. *Handbook of Oil and Gas Law*. Prentice Hall, Englewood Cliffs, N. J.

Chapter IX

Mining legislation in developing countries

Introduction

Economic conditions affecting petroleum exploration and production in developing countries quite naturally differ from those in industrialized countries. Exploration, development and production costs are normally higher, given comparable situations. Most of the equipment must be imported from the industrialized areas. Qualified personnel must be sought in North America or western Europe. Only as a developing country evolves into a major oil-exporting area, can equipment and personnel gradually be withdrawn from the area.

Local markets in developing countries are usually restricted, especially for natural gas. Oil or gas must therefore be transported for large distances before it reaches important markets. This leads on the average to higher costs and lower wellhead prices for oil and gas.

In addition to economic conditions, investing petroleum companies generally regard political conditions in developing countries as being unfavorable as well. Two types of political difficulties can be expected to arise. There is first the danger of major political disturbances such as riots or wars. Secondly, the companies fear eventual major changes in the political setting of a country that may result in basic changes in agreed contracts, or even nationalization.

There are many examples of the petroleum industry becoming the victim of riots and wars. The pipeline of the Iraq Petroleum Company (I.P.C.) to Haifa was closed in 1948 almost immediately after its completion and is still not in use. Two Suez-crises (1956 and 1967) disrupted the oil flow from the Middle East to western Europe. The civil war in Nigeria caused a disruption of oil production for SHELL-BP of more than one year; it also faced the companies with the dilemma of whether to choose the Lagos government, which could close the oil export point, Port Harcourt, with its small but effective fleet—or the rebel-government of Biafra, which possessed in the summer of 1967 almost all the oil fields of SHELL-BP in Nigeria (VAN MEURS, 1970).

Typical of political risk is that the chance of "something" happening to the investment because of political troubles increases with time. Major political disturbances can perhaps be anticipated in the years immediately following introduction of an investment. Political developments after a number of years, however, are often not possible to forecast.

The inclusion of political risk in the investment analysis is possible with the methods described in Chapter IV. One method is to use the pay-out-time yardstick, and to discount cash flow at an appropriate interest rate. Another technique is to incorporate

the political risk in the Monte-Carlo method through the use of random-fictive negative
outlays (or random "nationalizations") in the calculations. This analysis will result in a
lower expected-monetary value for a project.

Due to these political and economic drawbacks, the oil and gas projects that interest a petroleum company must incorporate a better outlook for richer fields than
similar projects in the industrialized countries (see also Chapter VI, p. 143).

As explained in Chapter III, the Middle East is fortunate in possessing a large
number of excellent oil fields. These fields are so prolific that they are a major destabi-
lizer of international oil trade. Much effort, therefore, has been spent by the major oil
companies in this area, with at present a large number of different contracts operating
between companies and governments. Since geological and technical conditions are comparable in the areas around the Persian Gulf, it is interesting to compare a number of
contracts within this Persian Gulf area. This will be the first and the most important
analysis in this chapter. A few observations concerning other areas will then follow.

Middle East

Conditions pressure of present contracts

Three different types of contracts are presently operating in the Middle East:
"equal-profit-split" contracts, participation contracts, and work contracts. These three
different types will each be analyzed in turn.

"Equal-profit-split" contracts

Most of the oil in the Middle East is produced under this type of contract. Included
are the contracts between N.I.O.C. and the Consortium in Iran, the I.P.C. in Iraq, the
contract with ARAMCO in Saudi Arabia, and with the K.O.C. in Kuwait.

Typical of all these contracts is the large area covered by the agreement. For
instance, the area granted to ARAMCO was originally 793,600 km^2, is now about
270,000 km^2, and will be at the date of expiration of the contract still 51,000 km^2.
Large areas are also assigned in the other contracts. Another aspect is the long duration of
the agreement. Table XIX demonstrates facts concerning these periods.

TABLE XIX

DURATION OF FOUR IMPORTANT MIDDLE EAST OIL CONCESSIONS

	effective date	duration	date of expiration
Consortium (Iran)	1954	40	1994
Iraq Petroleum Comp.	1925	75	2000
Kuwait Oil Comp.	1934	75	2009
ARAMCO	1933	60	1993

Originally, these contracts were interpreted with an equal-profit split, in such a way that the government and the company both earned 50% of the net profit. Royalties could be offset against income tax, and had the character of advance tax payments. The first contract of this type was concluded between ARAMCO and Saudi Arabia in 1950. The advantage of this regulation for the American Aramco-companies has been that tax paid to governments in foreign countries can be offset against domestic tax requirements in the U.S.A.

Before this type contract was established, companies were paying 4 sh per ton oil. Although this appears to be a minimal requirement, in the early 1930's this payment amounted to about half of the profit (MIKDASHI, 1966, p. 135). The new contracts drawn up in 1950 continued, therefore, an already established trend.

Apart from this 50/50 split, other financial requirements were sometimes applicable for particular contracts. For instance, the contract with the I.P.C. included a provision for a minimum tax payment of 25% of the value of the posted prices, or not less than £ 25,000,000.– (CATTAN, 1967, p. 49).

In the summer of 1962, O.P.E.C. adopted a resolution (Resolution IV. 33) in which the companies were required to regard the royalties no longer as "advance tax payments" but as compensation for the extraction of the petroleum. In most industrialized nations such a regulation has long been common practice. The companies accepted this principle after lengthy discussions. As a result, the original equal-profit split is now only applicable to the taxable income calculated after the expensing of the royalties. (MARTINEZ, 1966, p. 123).

The income in the Middle East countries is computed on the basis of posted prices. The realized price, however, is in a number of cases substantially lower than the posted price. The realized price was the posted price minus the discounts—ranging from U.S.$ 0.10 to U.S.$ 0.60. The posted price at Ras Tanura is U.S.$ 1.80 per barrel and the discounts are consequently a substantial part of the price.

The conditions pressure for these contracts can be calculated on the basis of either posted prices or realized prices. Fig. 42 illustrates the various conditions pressures, accounting for a varying cost rate and a varying amount of discounts on posted prices. The conditions pressure based on posted prices will be roughly 0.57; based on realized prices and reckoning with an average discount of $ 0.40, the conditions pressure will be about 0.75.

Due to the complexity of the world-oil market (see Chapter III) and the large-scale integration of the major international oil companies it is doubtful whether a conditions pressure based on the posted prices or the realized prices gives an adequate insight in the real "burden" placed on the major oil companies by the petroleum legislation in the Middle East. However, no other yardsticks are available. Since most authors believe that the posted prices for the exported oil (as $ 1.80 at Ras Tanura) are presently somewhat higher valued, it can be assumed that the "real" conditions pressure will be in any case not less than 0.57. On the other hand, very high discounts are exceptions and we can therefore postulate that the conditions pressure will seldom go above 0.8. Thus the present conditions pressure in the Middle East, given the present price structure in the

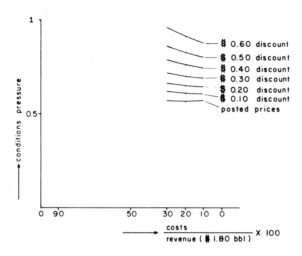

Fig. 42. Conditions pressures for O.P.E.C.-type Middle East contracts

world-oil market, can be narrowed down to between 0.6 and 0.8. Similar conclusions were given by CATTAN (1967, p. 79) and MIKDASHI (1966, p. 176), while the First National City Bank calculated about 0.68 (SHELL BRIEFING SERVICE, 1969).

Participation contracts

Participation contracts can take different forms. The most simple variation is government participation in the capital stock of an oil company. The first concession granted in the Middle East—the D'Arcy concession—included a participation in the capital stock by the Persian government. Similar participations have been realized by the government shares in BP (by Great Britain), in C.F.P. (by France) and in E.N.I. (by Italy). In this case the government has the role of a "sleeping landlord", which is not the type of contract that developing countries are seeking.

A special type of participation occurs when two governments agree to co-operate— such as the Franco-Algerian agreement for the exploration and production of oil in Algeria, or the agreement between Algeria and the Yemen-government. This type accord is typically bilateral; it has the disadvantage of an added risk, since a major disturbance in one of the participating countries may cause untold damage to the other partner. This type of participation agreement is therefore rare.

Two participation formula's are presently of interest:
a. participation of the government in an oil-production company;
b. participation of the government in an integrated oil company.

Participation in oil production. The first important contract of this kind in the Middle East was that between N.I.O.C. and A.G.I.P. in 1957 in Iran. The two companies formed a new one, S.I.R.I.P., that carried out on their behalf exploration, operation, and sale of crude and products, by providing each with 50% of the capital stock. A.G.I.P., however, was obligated to finance exploration with a minimum work obligation of U.S.$ 22,000,000.– with the stipulation that in case of the exploration's failure it would not be compensated. S.I.R.I.P. would finance the mentioned functions and would sell to A.G.I.P. approximately 50% of the oil produced.

The royalty amounted to 12.5% and the tax to 50%, calculated on the basis of posted prices. Originally the royalty could be offset against corporate-income tax; since the new O.P.E.C. agreement, however, the royalty had to be expensed. The conditions pressure of these terms is given in Fig. 43. Calculated on the basis of posted prices, the conditions pressure is about 0.8, on the basis of a realized price amounting to a 25% discount on posted prices, the conditions pressure is about 0.9.

A similar agreement was reached in Iran with Pan American, which together with N.I.O.C. formed the I.P.A.C.

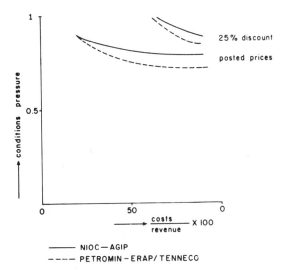

Fig. 43. Conditions pressures for participation contracts in the Middle East

Participation in integrated oil activities. An example of such a contract is that between PETROMIN and AUXIRAP. PETROMIN is the state-owned company in Saudi Arabia, and AUXIRAP is a company belonging to the E.R.A.P.-group. Later the American independent TENNECO took a share in the E.R.A.P. portion of the contract. AUXIRAP took the risk of exploration, while Saudi Arabia could participate for 40% in the integrated activities of the company if a commercial discovery was made. The royalty was determined according to a sliding scale: 15% as long as the annual production is below 60,000 barrel per day, and ranging to 20% above 80,000 barrel per day. The royalty was determined according to posted prices and had to be expensed. Taxation was 40% based on the posted price. For transportation there was a Saudi tanker preference.

Besides these regulations the contract included numerous details about refining and marketing. The conditions pressure with this contract is roughly 0.75 based on posted prices and 0.9 in the case of realized price 25% lower than the posted price.

In December 1967 a similar contract was closed between PETROMIN and E.N.I., in which the American independent PHILLIPS later took a share.

Work contracts

A completely different type of contract was concluded in Iran between N.I.O.C. and E.R.A.P. It departs essentially from the former types (PACHACHI, 1968) in establishing a new relation between the host country and the company. The latter has in work contracts only the function of rendering services to the host government. In exchange for the outlays for exploration, the company obtains the right to purchase an established amount of oil when the exploration results in a discovery. In the case of a failure, the service-company bears the entire risk. E.R.A.P., in this contract concluded in 1966, is regarded as a non-profit organization. It is entitled to purchase between 35 and 45% of the half of the discovered reserve. The other half is set aside as "national reserve". In the case of a discovery the exploration outlays are repaid by N.I.O.C. The price to be paid is costs plus 50% of the difference between costs and realized prices. Since E.R.A.P. acts as a non-profit organization, no royalty need to be paid.

Special provisions are included in the contract such as the stipulation that an eventual marketing by E.R.A.P. or N.I.O.C. oil will be linked with purchases of French goods by Iran.

The conditions pressure that originates from this contract depends upon many complexities rather difficult to pin down. A rather intensive discussion about this subject was opened by STAUFFER (1967). A superficial analysis of the E.R.A.P. contract would lead to the assumption that the profit split would be about 90/10 to the advantage of the Iranians—these figures are frequently quoted. This 10% originates from the fact that Iran has the right to set aside 50% of the reserves; from the remaining 50% about 40% can be bought by E.R.A.P., which is 20% of the total discoveries; at the same time E.R.A.P. is paying 50% tax on these sales, reducing the profit for E.R.A.P. to 10%. STAUFFER (1967), however, argues rightly that this is not a proper comparison. Since the oil belongs entirely to N.I.O.C., it can sell its own oil only for realized prices instead of posted prices. Further, a highly advantageous compensation to E.R.A.P. for the exploration and developments costs is included in the contract. The result is that N.I.O.C. will not generate profits for a number of initial years, while E.R.A.P. is directly realizing its advantage of earning 50% of the difference between the costs and realized price. STAUFFER (1967) therefore concludes that the deal is ultimately no better for Iran than a normal O.P.E.C.-type agreement. A number of details were challenged in the note by SHAIR (1967) but this author also admitted that the 90/10 split was fictitious.

Although it is with this type of contract almost impossible to speak of a real conditions pressure, it can be assumed that the E.R.A.P. risk in this deal was not larger than the normal O.P.E.C. agreements.

In addition to the singular character of the contract, its duration is relatively short—only 25 years. This leaves N.I.O.C. with what is left of the reserve after 25 years.

Further, it must be mentioned that realized prices may change. An upward movement of the realized price would make the contract more favorable to N.I.O.C.

A similar contract was concluded between the Iraq National Oil Company and E.R.A.P.

Comparison between the contracts

An important difference in conditions pressure exists between the "equal-profit-split" contracts and the participation contracts. The conditions pressure of the former ranges from 0.6 to 0.8 while the conditions pressure of the latter will range from 0.75 (0.8) to 0.9.

It is logical that the companies in the Middle East accepted higher terms for the newer contracts concluded after 1957, since more of the geology was known and economic and technical conditions were promising. Originally—in the 1930's—the conditions pressure for the concessions around the Persian Gulf, was about 0.5. Due to the equal-profit-split contracts concluded in the early 1950's, this conditions pressure was reaffirmed. The fixation of posted prices and the obligation to calculate tax after posted prices resulted in a gradual increase of the conditions pressure. Finally, the requirement to expense royalties brought the level between 0.6 and 0.8.

The new contracts fall, with regard to the conclusions of Chapter VI, in the range of a high probability-of-success ($0.25 < p < 0.5$). It is, however, questionable whether the ranges mentioned in Chapter VI can be used here, since these ranges were based on expected outcomes from 5 to 500 times the exploration effort (see Fig. 38). Due to the special characteristics of the Middle East, the expected outcomes are mainly concentrated in the upper part of the rectangles. Or in other words, the companies may have accepted the terms on the basis of a moderate probability-of-success ($0.1 < p < 0.25$) and a large average-expected profit.

Although the situation in the Middle East is politically rather explosive, the high conditions pressures accepted by A.G.I.P., E.R.A.P. and the American independents indicate that this risk plays a rather limited role in their calculations. The reason for this may be that pay-out time is generally short when a Middle East-type discovery is made. Further, the acceptance of these terms will limit the possibilities for major changes in mining legislation in these countries, since very little oil revenue remains to be earned from the companies. A major attack on the A.G.I.P. and AUXIRAP agreements is rather pointless.

The acceptance of the A.G.I.P. and AUXIRAP contracts raises the interesting question of whether the "equal-profit-split" contracts are out of date. Due to the lengthy duration of these contracts (see Table XIX), the original risk taking by the companies is extensively honored by large revenues. Therefore, the companies are presently earning large true rents from these contracts. This is a logical argument for renegociation of the present "equal-profit-split" contracts, especially if new economic and technical developments are considered (see also Chapter III).

The introduction of the work contracts initiated a new period in the mining legislation in the Middle East. It is, however, questionable whether the purely financial benefits to the government are as large as with the participation contracts. The character of the new agreements is, however, for a large part also a political choice.

Other areas

The areas around the Persian Gulf are characterized by their highly prolific oil production. With the exception of Lybia, no other areas in the world show a production as prolific as that in the Middle East. The question may arise, therefore, whether conditions pressures equal to those in Middle East contracts (0.8 to 0.9) can be negotiated in other areas.

A uniform petroleum law seems in any case easy to handle and would lead to an optimum allocation of investments by the oil companies in the different countries. The idea to work for a uniform petroleum code was articulated in 1964 by O.P.E.C. (resolution V.41). The need for uniformity was expressed earlier in 1958 in the explanatory statement accompanying the Sahara Petroleum Code. According to CATTAN (1967, p. 22) the statement said: "It appears opportune to make the regime for petroleum operations uniform throughout the Sahara which constitutes an original and homogeneous geographical entity . . . It would indeed be difficult to envisage that enterprises operating under the same technical, climatic and geographical conditions should be subject to different juridical and fiscal regimes which could be the source of artificial distortions of competition and could also be prejudicial to the common good."

Although the previous statement and the ideas of the O.P.E.C. seem logical, an important reservation must be noted. Even amoung the O.P.E.C. countries, geological conditions controlling oil exploration and production differ widely. This can be illustrated with figures for the daily production per producing well. In Iraq and Iran the daily production is about 14,000 bbl per producing well, while in Indonesia and Venezuela this figure is about 350 bbl. The productivity in the Middle East per well is consequently 40 times larger than that in Indonesia or Venezuela (see also Fig. 21).

From the analysis in Chapter VI it can be concluded that different geological conditions require different conditions pressures, if a country wishes to maximize the public revenue from the operations. Different geological conditions should necessarily result in different petroleum laws if the maximation of the public revenue is the target of the legislation. A uniform petroleum law is therefore not in the interest of the majority of countries. Conditions in Algeria are different from those in Iran and a different petroleum law must therefore be constructed to meet different needs. A uniform petroleum law would lead to a situation where a number of countries (for instance those of the Persian Gulf) would grant a large rent to the companies, while the economically-recoverable reserve in other countries would become too small to induce companies to conduct sufficient exploration.

There is no question of a distortion of competition if countries that are able to earn a large rent effectuate this with firm terms, while countries that are able to earn only a small rent consequently work with more attractive terms.

It is therefore not in the interest of the developing countries to work for a uniform petroleum code, even on a regional scale. Countries with geological conditions less favorable than in the Middle East should offer the companies more favorable terms.

Conclusions

The accent of the Middle East petroleum laws is on taxation and state participation. This accent is in line with the selectivity principle outlined in Chapter VI. It may therefore be concluded that the recent participation contracts in the Middle East are well suited to reaching the maximum public revenue. The O.P.E.C.-type contracts, under which most of the present Middle East oil is produced, seem outdated. It must be mentioned that a tendency towards a uniform petroleum law for the O.P.E.C.-countries, even in the Middle East and North African oil-producing countries, is not in the interest of the various governments.

Literature

References

Cattan, H., 1967. *The evolution of oil concessions in the Middle East and North Africa.* Publ. for the Parker School of Foreign and Comparitive Law. Oceana Publications, Inc. Dobbs Ferry, New York, 174 pp.

Martinez, A. R., 1966. *Our gift, Our oil.* NV. D. Reidel, Dordrecht, 199 pp.

Mikdashi, Z., 1966. *A financial analysis of Middle Eastern Oil Concessions: 1901-1965.* Preager, New York, 340 pp.

Pachachi, N., 1968. Disparity of Concession terms in Middle East Oil Producing Countries. *Third Seminar on Petroleum Economics and Development.* Kuwait Institute of Economic and Social Planning in the Middle East, 23 pp.

Shair, K., 1967. The ERAP Agreement: the discounted cash flow and mr. Stauffer's analysis. *Middle East Economic Survey*, X(27): 1-10, 5 May 1967

SHELL BRIEFING SERVICE, 1969. The economic impact of oil operations in producing countries. Shell Briefing Service 9 p.

Stauffer, T. R., 1967. The ERAP agreement: a study in Marginal Taxation Pricing. *Sixth Arab Petroleum Congress*, Bagdad, 6-13 March, Paper 72 (A-1).

Van Meurs, A. P. H., 1970. Olie en Biafra. *Econ. Statist. Ber.*, 55 (2729): 56-60.

Selected documents

Alnasrawi, A., 1967. *Financing Economic Development in Iraq—The Role of Oil in a Middle Eastern Economy.* Praeger, New York, 188 pp.

Hatry, P., 1967. La Contribution de l'Industrie Pétrolière á la Balance des Paiements des Pays en Voie de Développement de 1966 à 1975. *Cahiers Economiques de Bruxelles*, 33: 5-62. Dep. d'Economie appliqué de l'université libre de Bruxelles.

Leeman, W. A., 1962. *The Price of Middle East Oil. An Essay in Political Economy*. Ithaca, Cornell University Press, 274 pp.

Longrigg, S. H., 1961. *Oil in the Middle East. Its Discovery and Development.* ISSUED UNDER THE AUSPICES OF THE ROYAL INST. OF INTERNATIONAL AFFAIRS. Oxford University Press, London, 401 pp.

Penrose, E. T., 1968. *The Large International Firm in Developing Countries—The international Petroleum Industry*.

Tanzer, M., 1969. *The Political Economy of International Oil and the Underdeveloped Countries*. Beacon Press, U.S.A., 435 pp.

Conclusions

To realize the maximum public revenue from activities of petroleum companies a number of financial arrangements must be included in the petroleum law influencing these activities. This law should be selective and dynamic.

Selectivity implies that the terms should be regulated according to the ultimate profitability of the fields. In small or developing countries the selectivity can be achieved by giving weight to elements as corporate-income tax, state participation and sliding-scale royalties. The terms must be adjusted to geological, economic and technical conditions.

A dynamic petroleum law must include the possibility for renegociating the contract if economic and technical conditions change radically, or, if companies have been overcompensated for the original taking of the enterpreneural risk.

This study concludes that the intensive reformulation of the Dutch offshore-mining legislation during 1966 and 1967 caused practically no change in the conditions pressure.

Almost no differences in conditions pressure exists between the west European and American laws affecting oil production. For natural gas the conditions pressure in the Dutch legislation is considerably higher but still consistent with a sound mining policy.

The tendency to work for uniform financial arrangements for petroleum production in the developing countries is not in the interest of these states, because it is contrary to the selectivity principle. The O.P.E.C.-type agreements in the Middle East should be renegociated.

Appendix I

Hoskold's formula
The derivation of Hoskold's formula, is given for constant yearly earnings from a mining property, beginning a year after the investment.

Let:

V	= present value,
r'	= speculative rate to purchaser on his capital investment,
r	= practicable safe rate on redemption of capital,
E	= yearly earnings to be purchased,
n	= life of the project,
A	= amount of an annuity of U.S.\$ 1.— per year for n years at rate r.

Then:

$E - V r'$ = annual redemption, because $V r'$ is the expected interest on capital investment.

$(E - V r') A$ = the total redemption amount.

This amount must be equal to the purchase price for the property of V.

Since the annuity is $\dfrac{(1+r)^n - 1}{r}$, the present value of the property can be expressed as $V = \dfrac{E}{\dfrac{r}{(1+r)^n - 1} + r'}$

This last formula is called Hoskold's formula.

An example:
What is the present value of a mine that gives U.S.\$ 10,000.— earnings over three years, while the redemption of capital can be reinvested at a rate of 4% and the investor wants a speculative interest rate of 10% on invested capital?
Hoskold's formula gives as a result: V = U.S.\$ 23,790.—
This amount can be recovered as follows:

year 1:	earnings	U.S. \$ 10,000.—
	speculative interest	U.S. \$ 2,379.— —(10% of \$ 23,790)
		U.S. \$ 7,621.—

U.S. \$ 7,621.- increases in value in 2 years at an interest of 4% to:		U.S. \$ 8,236.—
year 2: gives also U.S. \$ 7,621,- which increases in 1 year to:		U.S. \$ 7,926.—
year 3: gives also:		U.S. \$ 7,621.—
		U.S. \$ 23,783.—

LITERATURE

Hoskold, H. D., 1889. Memoire Général et Spécial sur les Mines, la Metallurgie, les Lois sur les Mines, les Resources, les Avantages etc., de l'Exploitation des Mines dans la République Argentine, Imprimerie et Stéréotypie du "Courrier de la Plata", 832 Mexico y Bolivar 230; 620 pp.
Parks, R. D., 1949. Examination and Valuation of Mineral Property. Addison-Wesley Press Inc., Cambridge, Massachusetts, 504 pp.
Frick's Petroleum Production Handbook, 1962., Vol II, McGraw Hill, Chapter 38.

Appendix II

Arps' formula
The derivation of Arps' formula is described by Arps in the following way (Arps, 1958, p. 339): "The weighted average deferment factor (defined as the ratio of the discounted value of the future production increments to the undiscounted total future production) for r per cent interest per year on the production and, therefore, also on the total operating net revenue ($E = \Sigma\ e$) will be designated as D. Since the amortization is on a unit-of-production basis, the same average deferment factor, D, applies to the total of these amortization payments ($C = \Sigma a$). If the drilling investment, C, had been borrowed from a bank it would be obvious that this indebtedness could be discharged by returning each year the annual amortization payments, a, to the bank plus an annual interest payment of $\frac{rB}{100}$ at the bank's rate of r per cent on the remaining balances, B.
The present value of the combined amortization payments, $D\ \Sigma\ a$, plus the present value of the interest payments, $\frac{rD'}{100}\ \Sigma\ B$, is therefore equal to the total original investment, C, regardless of the amortization pattern used.

$$C = D\ \Sigma\ a + \frac{rD'}{100}\ \Sigma\ B$$

in which D is the weighted average deferment factor on the production of the well, the total earnings and the combined amortization payments, while D' is the weighted average deferment factor on the undepreciated balances, B, of the investment. Both deferment factors are computed at the same safe interest rate, r, compounded annually. Since

$\Sigma\ a = C$, it follows that: C

$$(1-D) = \frac{rD'}{100}\ \Sigma B$$

and the present value of 1 per cent interest on the undepreciated balances is:

$$\frac{D'\ \Sigma\ B}{100} = \frac{C\ (1-D)}{r}$$

Some twenty years ago the engineering staff of the Shell Oil Company first combined this expression with the present value of net profits after amortization:

$$D\ (\Sigma e - \Sigma a) = D\ (E - C)$$

and demonstrated that if the present value of 1 per cent interest on the undepreciated balances of the investment equals $\frac{C\ (1 - D)}{r}$, then the present value of the net profits after amortization, $D\ (E - C)$, must be equivalent to an earning power or average annual rate of return r, expressed in per cent of the undepreciated balances of the investment of:

$$r' = \frac{rD\ (E-C)}{(1-D)C} \quad \text{per cent per year}$$

This very useful and logical relationschip represents the equivalent annual rate of return on the undepreciated balances of the investment, provided the investment is amortized on a unit-of-production basis".

LITERATURE:

Arps, J. J.; 1958. Profitability of capital expenditures for development drilling and producing property appraisal. Transactions AIME 1958, Vol. 213, pp. 357-344.

Frick's Petroleum Production Handbook, 1962, McGraw Hill, Chapter 38.

Appendix III

The portfolio-problem

Three investments are influenced by the same events, but the profitability of each of these investments is influenced in a entirely different way. The investments are:

(1) A1, an exploration venture,
(2) A2, a similar exploration venture,
(3) B,   a secondary-recovery project.

The two events are:

(1) $e1$ the absence of oil in the untested basin.
(2) $e2$ the presence of oil in the untested basin.

The net-present values of the three projects for two different outcomes are given below:

event A1 (expl. project)	A2 (expl. project)	B (sec. rec. project)
$e1$ − U.S.$ 10.− MM	−U.S.$ 10.− MM	+ U.S.$ 11.− MM
$e2$ + U.S.$ 190.− MM	+ U.S.$ 190.− MM	+ U.S.$ 1.− MM

If event $e1$ has a possibility of 0.9 of occurring and event $e2$, a possibility of 0.1, then the expected-monetary value of the three projects can be given as:

event	probability	A1 and A2	B
$e1$	0.9	−U.S.$ 9.−	+U.S.$ 9.9
$e2$	0.1	+ U.S.$ 19.−	+U.S.$ 0.1
expected-monetary value		+ U.S.$ 10.−	+U.S.$ 10.−

The expected-monetary value of two different combinations of two projects is:

	A1 (or A2) +B	A1 + A2
$e1$	− U.S.$ 9.− + $ 9.9 = U.S.$ 0.9	2 x − U.S.$ 9.− = − U.S.$ 18.−
$e2$	+ U.S.$ 19.− + $ 0.1 = U.S.$ 19.1	2 x + U.S.$ 19.− = + U.S.$ 38.−
	U.S.$ 20.0	$ 20.−

Expected-monetary value of the combination of two projects.

It can be seen from this example that a certain combination (in this case A1 or A2 + B) of projects considerably reduce the exposure to risk.

Appendix IV

Classification of Petroleum

No uniformity exists at the present time in the nomenclature used for the different kinds of petroleum reserves. The following definitions are recommended:

Petroleum. A gaseous, liquid, semisolid or solid mixture of hydrocarbons and hydrocarbon compounds occurring naturally in the rocks.

In petroleum production practice only those hydrocarbons or hydrocarbon compounds normally produced through wells are considered. When recovered, they may be in the liquid phase (liquid hydrocarbons) or in the gaseous phase (natural hydrocarbon gases).

Liquid Hydrocarbons. These are subdivided into:

(1) Crude oil, consisting primarily of intermediate and heavy hydrocarbons and hydrocarbon compounds and occurring in a natural-liquid state under reservoir conditions. They are produced and recovered as a liquid by ordinary field separating equipment.

(2) Natural-gas liquids, liquid hydrocarbons consisting primarily of light and intermediary hydrocarbons and occurring as a free-gas phase or in solution with the crude oil in an oil reservoir. They are recoverable as liquids by the processes of condensation or absorption in field separators, scrubbers, gasoline plants, or cycling plants. Based on differences in recovery and storage methods natural-gas liquids are subdivided further into:

    (a) Condensate, consisting primarily of low-vapor-pressure products recoverable by ordinary field separator equipment in the same manner as crude oil. For practical purposes such condensate, if actually recovered on the lease by standard field separator equipment, is often combined and recorded together with crude oil;

    (b) natural gasoline, consisting primarily of intermediate-vapor-pressure products, which are recoverable by special equipment or gasoline plants;

    (c) liquefied-petroleum gases, consisting of high-vapor-pressure products such as butane, propane, and ethane, which are recoverable in specially-equipped gasoline plants and which can be maintained in the liquid phase only under substantial pressure.

Natural Hydrocarbon Gases. Natural hydrocarbon gases primarily consist of the lighter paraffin hydrocarbons and are subdivided into:

(1) non-associated gas, gaseous hydrocarbons occurring as a free-gas phase under original conditions in a reservoir, which is not commercially productive of crude oil;

(2) associated gas, gaseous hydrocarbons occurring as a free gas phase under original reservoir conditions in contact with a commercially productive crude-oil reservoir;

(3) dissolved or solution gas, gaseous hydrocarbons, occurring under original reservoir conditions in solution with crude oil in a commercially productive crude-oil reservoir;

(4) injected gas, gaseous hydrocarbons which have been injected in underground reservoirs for pressure maintenance or storage purposes.

LITERATURE

Frick's Petroleum Production Handbook, 1962, McGraw Hill, Vol. II, Chapter 37,pp. 37-6 and 37-8.
        37, pp 37-6 and 37-8.

Appendix V

Conversion factors (approximate)

1 foot (12 inches)	= 0.3408 m
1 metre	= 3.281 ft.
1 mile	= 1.609 km
1 square mile	= 640 acres = 2.59 km^2
1 acre	= 4,047 m^2
1 American barrel (bbl)	= 42 American gallons = 159 l
1 barrel oil per day (bopd)	= approx. 50 tons a year (depending on weight of crude)
1 cubic foot	= 0.0283 m^3
1 m^3	= 35.3 cuft. = 6.29 bbls
1 therm	= 100,000 B. Th. U. = 25,200 kcal.
1,000 kcal.	= 1 thermie = 0.03968 therms = 3,968 B. Th. U.